The History of Naval Medicine

Cover Art – from the US Navy Medical website. One old saying was that once upon a time "The men were made of steel and the ships were made of wood" by a male Lt. Now that's the real Old Navy. Another saying that was significant was from a female Lt Cdr in charge of ADP operations – "Lead, Follow, or Get out of the Way". In my travels I met the man who was the Executive Officer in charge of turning the USS Constitution into a floating museum. The ship on the cover is one of these revolutionary period sailing ships. This so called gentleman was a Naval Academy graduate and worked at a local company on a project with me. He now works at IBM research. He was one of the few supervisors I did not like from the Navy due to his style of management – overbearing theory X with me doing all his dirty work – not a leader by example.

Forward

This book was written for the people who work at Navy Medical Command who are dedicated and never ask for anything other than interesting work to do and to be able to help others health care in service to their country. It takes a special type of warrior class to care for others when they are most vulnerable. I made many friends and a few enemies at Navy Medical and often said I would write a book about my experiences there one day. Well, this is a book that is more about the history, evolution, and development of Navy Medicine since the beginning of time.

I do not believe a book of this nature has been done at this time in history and believe it will enlighten those who wish to undertake a career in Navy Medicine whether they are civilian or military. Most of the information used in this text was collected via the internet or from my memories of Navy Medical. This is not a sanctioned official history of Navy Medicine but my unbiased viewpoint written after 10 years of civil service at Bethesda realizing the fact that I learned many good things while I was employed by Navy Medical and use those things every day in my current job.

God bless those at Bethesda Navy Medical Center and BUMED in Washington DC, my first true work home from 1977-1987. I would also like to thank my children, loving wife, parents, siblings, relatives, friends, coworkers, staff, and pastors for keeping the faith in me all these years. I've made it a more

difficult trip than it had to be but my heart was in the right place. I am thankful that

Navy Medical gave me my second start in government service as a student

intern after the Air Force Academy. This book is unclassified and any sensitive

information has been deleted or left out intentionally. It was written in good faith

to keep a promise to my coworkers to tell their story of suffering and plight at

Navy Medical Data Services Center, Bethesda, Maryland.

Table of Contents

List of Figures

1. Wars without Death

Let's talk about a novel idea. Suppose two world leaders get into an argument and decide to go to war over the issue. Neither of them have children in the armed forces of their countries so their decision to go to war does not directly impact their family. They both push a button on a computer war game and a number of statistical casualties on both sides are subtracted from their general population. Or better yet only a large sum of money is subtracted from their national financial accounts and the loser is the one who loses the most money. This would be war without real death and is actually a borrowed idea from an old Star Trek episode. Computers have this type of processing power today yet can not actually kill anyone. It would be correct to say that computer power and technology does kill people, though, through our advanced weapons systems. The generation of baby boomers has lived with the threat of nuclear holocaust our entire lives. For us there has been no easy sleeping with world leaders at the helm of a potential global disaster. Older generation talk about a time when people where more kind, yet in was the older generation who brought the sins of atomic power to bear in World War II in order to defeat an enemy who was predisposed to sacrificing everything to win. Since those times the public has demanded that we curb the national desire for dominating the world and we call this nuclear détente or mutual destruction. With 17 nations in the nuclear club according to the Federation of American Scientists, management of the fear of nuclear weapons has never been more important. Once a country has this capability it can dictate demands to the developed countries of the world. In deed, there is a view that the under developed countries are the ones that need nuclear weapons the most.

Until we reach a point where wars can be fought without death as in this scenario, we will always need a Navy Medicine group to help treat the wounded soldiers of our national conflicts. The people of Navy Medicine realize this mission is important and are some of the nicest warriors on the planet. Inside I believe most of them understand and despise wars and the damage they can do to families. When someone asks if the military is always a brutal adventure one can honestly say that Navy Medicine is the antidote to a lot of the ill perceptions of the military. In fact, many doctors, nurses and dentists get their start at Navy Medical and the other military branches before they go into private practice at a civilian hospital. Combat surgery has been a major way that doctors get experience in an emergency triage situation. This was very well documented in humor though the TV series called MASH – Mobile Army Surgical Hospital. Mobile hospitals get the doctors to where they are needed in the field. What better way to do this than with a giant hospital ship? This book is an attempt to understand the history of global Navy Medicine and the politics that helped create the US Navy Medical into one of the greatest health care systems on the planet.

2. Ancient Greek Medicine

Greece was the first culture to practice medicine in a scientific way. We get the Hippocratic Oath from the Greeks which states that the doctor's first loyalty is to curing the patient through clinical observation. They treated people with a system of medicine just as they used systems thinking in philosophy, art, mathematics, religion, and other westernized academic disciplines. There was a Greek Navy, therefore one can correctly assume there were Greek Navy doctors. At the battle of Troy which was won mainly while at sea on the shores of the Trojans in current day Turkey, we can surmise that battle casualties were treated by doctors who traveled by ship to Troy with the kings and men who made up the Greek army. Greek doctors and medicine relied on herbs and plants to cure patients. The main battle wounds of the Greeks would have been from crude weapons of the bronze age. The Myrmidons, Achilles elite brave soldiers stormed the shore at Troy under fire using the turtle of shields to protect themselves from arrows as depicted in the movie called "Troy". The US Military academy has verified that the order of battle followed the movie closely with computer simulations. This qualifies as an amphibious naval engagement of Greeks storming the beachhead and capturing the nearby Temple of Apollo. Trojans worshiped Apollo and the God's of the Greeks and later founded Rome (Aneus in the Aneid). The unfortunate thing is that Homer and Virgil did not write their poems of the battle of Troy until many years (500) after the event and they did not capture the medical attention of the war. They did capture the fact that the Greeks felt a deep kinship to their fallen comrades and gave them a decent burial through cremation. Indeed, Navy medicine and military medicine is often the red headed step child of the military where people who care about health were once singled out as "weak". Yet, the Greeks truly understood that healthy fighters were able to build their health through the Olympic games and better health care practices in their daily lives. In 2004, the Olympic games returned to Greece and the greatness of their history came to light in the American public and world. The people and Armies of Greece were known by the cities they fought for. Athenians were known as academics and fighters. Spartans were known as the best military warriors to the point of cruelty. There were indeed many civil wars within Greece itself for control of the country by the Kings.

Some of the famous Greek doctors were Hippocrates[1] and Galen. Hippocrates was practicing 430 years before Christ and well after the Trojan war. He practiced the idea of clinical observation whereby a doctor would view a patient for a period of time while he was sick. He wrote 60 volumes of medical information. He was known as the "Father of Medicine". Galen was a doctor who practiced in later day Rome who was a Greek. He was allowed to practice medicine under Caesar. They both practiced medicine for their kings with the idea that they could lengthen life and please the many Gods of Greek mythology of the time.

---

[1] Hippo crates, Accessed from internet webpage www.historylearningsite.co.uk/hippocrates.htm

It has been said that Jesus was a Hellenic Jew[2] which means he understood many of the customs of the Greeks and the worshipping of multiple Gods. This may have contributed to his understanding of their advanced culture and views of health and medicine. Jesus was known as a healer and he used his faith to heal people. He surely knew about the herbs and medicines that the Greeks used in addition to the systematic ways they developed to observe patients and heal them. One United Methodist pastor said that all western culture gets a lot a circular thinking and systematic developments from the Greeks including a Greek version of the bible and the Coptic gospels. Surely, through Jesus and the Hellenic Jews we have continued the Hippocratic Oath and a "First Cause no Harm" philosophy. Jesus actually said "Doctor heal thyself first". The Gospel of Luke, himself a doctor, discusses in great detail all of Jesus's great works including the faith healings. Interestingly, Luke was not one of the 12 apostles but his Gospel is included right there in the bible.

---

[2] The Bloodline of The Holy Grail: The Lineage of Jesus.

3. Ancient Roman Medicine

The Romans practiced medicine as a matter of daily life. They had Roman baths as far west as Britain for daily cleansing. Galen lived and practiced in Rome. He was a Greek but he practiced for the Caesar at the time. The Romans also had a Navy of some power. They held mock Navy battles in the coliseum when they flooded it with water. The viaducts channeled water to all parts of Rome ensuring good hygiene and Roman civil engineering projects. The Roman army was required to perform exercise as a daily routine and stay on a specific diet. These were the first controls on an Army of this sort. After all, Rome could not have survived without a healthy Roman Legion. Rome was the center of the western world for so many years it was a gift to be a Roman citizen. The society of Rome flourished with new engineering and sciences. Roman ships were built to carry them across the Mediterranean Sea and to many places in the provinces. Slaves were used to row the oars. Keeping the slaves healthy would have meant that no Roman would row a boat so it was in their best interest. The Romans practiced good health and hygiene as a part of daily life when they built their sewers system. Military hospitals had sewer systems and toilets in them as did all the homes. A healthy soldier meant a healthy empire. The system was designed for rich and poor people making it one of the first public health programs for all in the world.

Romans learned a lot from Ancient Egyptians. This would have been at a height when Mark Anthony was in love with Cleopatra. Both would have had access to the doctors of Egypt as Royalty and they certainly planned to bring this to the Roman Empire if they had won the war with Octavius. Gibbons book series "The Fall and Decline of the Roman Empire" chronicles the times that these people lived during in Rome. As the Roman Empire spread to Gaul and the provinces, the practices of bathing were spread too. Bath, England has some of the Roman type public baths that were designed for public health. These can still be viewed in this English town. Thus, the Egyptian system of medicine influenced Rome and Rome influenced Britain and the western world of today.

4. Ancient Egyptian Medicine

The ancient Egyptians were some of the most scientific of cultures. They were great engineers and doctors. They knew how to embalm their kings in a sarcophagus and mummify a human body. This took extreme knowledge of the human body and how to preserve it. The had a strong believe in the forces of nature that make life possible and worshipped the sun god, Ra. The pyramids stand as a lasting memorial to their kings and abilities to lead the ancient world in the building and engineering sciences as well as medicine. They invented writing on papyrus and encryption coded messages. To say they were good at mathematics is an understatement. They were the center of the world in the middle east during old testament times. The Jewish people who were enslaved by Egypt learned their medicine ways and transported them to the new lands of Exodus. Doctors were specialists in each part of the body and Egypt had many of them. They were servants of both royalty as well as the common people.

Egyptians mastered the medical field by experimenting with mummification methods and using instruments to practice a clinical style of medicine. One of the papyrus's preserved for mankind was a list of medical treatments for various types of ailments. This list covered head and neck injuries and is one of the oldest (BC) lists of medical treatments of it's mind in the world. The treatments are based on clinical observations of the patients and trial and error. Those treatments that are most likely to help the patient are outlined as being for use by the doctor. Those that were untreatable were stated as such. A case of medical tools was uncovered in one archaeological site. The chest had instruments of all types for the patients as well as a sack of herbs and plants. Because Egypt was so fertile near the Nile River, many plants grew there that did not exist in the rest of the world at that time. These plants had many healing properties and were useful in treating patients with herbal medicines. These spread to other parts of the world. Today, herbal; remedies are still used to help people in certain situations. Homeopathic remedies base in large part upon the use of remedies that are natural and found in the environment and not manufactured. The Egyptians may have qualified as the world's first pharmacists due to their large variety of plants.

The Egyptians also had doctors who specialized in keeping the armies and military healthy. It is suspected that Ancient Roman and Greek doctors learned what they could from the Egyptians and used their techniques in the rest of that part of the world. It would be interesting to do a comparative study of Asian cultures to find if Egyptian medicine methods found their way to that part of the world also. Certainly global trade merchants would have had the ability to import medical treatments from all parts of the ancient world via water travel and trade agreements. The fortunate thing about the Egyptians was that they were so very interested in preserving the dead that they also preserved many artifacts in Pyramids with the bodies of their dead. It is because we can examine these artifacts today that we have a better understanding of exactly just how advanced they were in medicine.

The Egyptian navy would have had military doctors from influential families whom specialized in various types of medicines. There is evidence that the Egyptian doctors were the first ones to specialize from the recorded observations and remedies for various ailments that are organized by the parts of the body. It is not clear if there was any combat medicine or trauma medicine practiced at that time with triage priorities. Instead they seem to have observed the patients one at a time and recorded observations and remedies that were successful and unsuccessful on papyrus for posterity. Since the Egyptian coastline and Nile River are basic waterways we can assume they had a powerful navy at the time of Moses. The river was also used to carry barges full of stone to the pyramids. Their engineering feats required healthy people to perform the work which would have given rise to the doctors helping the normal people in the society. The workforce had to be well maintained to work hard.

5. Royal Navy Medicine

        The movie "Master and Commander; the Far Side of the World" showed the relationship between a Royal Navy ship's captain and his chief surgeon in the 1700's during the period of enlightenment in Scotland. The University of Edinburgh medical school is indeed at least this old. The surgeon was concerned all about the health of the men, science and discovery on the trip and the ship's captain was concerned about the overall crew including the surgeon's concerns. The ship's captain could in no way perform the operations the surgeon could perform nor was he qualified in the medicine and sciences as the surgeon was. If the surgeon was injured he was replaced by his first surgeon mate. This was the Royal Navy at it's height of power. Discoveries in science were made by traveling around the world by sea. The Beagle was the British ship that sailed to the Galapagos island with the scientist Charles Darwin when he made his discoveries and wrote the book entitled <u>Natural Selection: The Origin Of Species</u>. Darwin was a college drop out who hated organized religion and sailed to escape the religious education system of the time. Some of this fact are seen in this movie with the surgeon looking for making discoveries in foreign lands.

        Like it or not, the American Navy was a combination of what came before us – the French Navy whom helped America at Yorktown and The

Royal Navy thus navy medicine evolved from what we knew about the navy medicine in these navies of the world. And it was Benjamin Franklin who traded electricity to the French for the help of the French fleet at Yorktown – a little know historic fact. Today, there is cooperation among the free countries navies of the world for the good of the medical patient but this was not always the case. The Royal Navy webpage shows an organization that is dedicated to controlling disease in the military services. Appendix C is a reading list of Royal Navy Medicine books from the Imperial War Museum Library of Great Britain.

The Royal Navy currently maintains a Royal Naval Institute of Medicine as a shore activity. Appendix D shows the webpage for this organization and some of the concerns of current Naval Medicine in all navies of the world. It is important to understand that many of the world's navy medical departments are very similar in function, problems, and way they conduct business. Figure 1 shows the Royal Navy Occupations and they are very similar to the US Navy Medical Department (BUMED).

Figure 1. Royal Navy Occupations Matrix

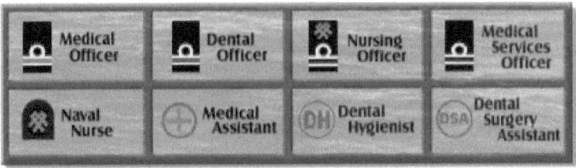

Figure 2 shows the scope of the Royal Navy Medical Service and the various functions they perform in today's terms.

Figure 2. Scope of the Royal Naval Medical Service:

- Primary Care
    - General Practice
    - Occupational Medicine
        - General OM
        - Aviation Medicine
        - Submarine and Radiation Medicine
        - Diving and Hyperbaric Medicine
    - Public Health
        - Environmental Health Inspectors
- Secondary Care
    - Medicine
        - General
        - Paediatrics
        - Psychiatry
        - Haematology
    - Surgery
        - General Surgery
        - Orthopaedic Surgery
        - Urology
        - Maxillofacial

- o Anaesthetics

- o Pathology

    - Histopathology

    - Microbiology

    - Biochemistry

- Dental Surgery

- Nursing

- Paramedics

- Medical Administration

Appendix E shows the statement of the role of Royal Navy Medicine in Great
Britain's Navy from their current webpage. The parallels between the Royal;
Navy and the US Navy medical departments are uncanny. One must remember
that the Royal navy ruled the seas for many years and Great Britain is still a
country surrounded by water whom depends on the sea for trade, food, fossil
fuel, and transport. They remained a powerful navy well into the second world
war.

6. Revolutionary War

John Paul Jones lies entombed at the US Naval Academy Chapel. He could not have envisioned how global and sophisticated the modern navy would become. The modern Naval Medical Command has Medical treatment facilities on almost every navy base in the world. After ten years of working with their data processing group, there was always plenty of data on the worldwide operations, management, and budget of Navy medicine. The revolutionary war days had to have been mighty hard on the sailors and surgeons assigned to the ships. The lower decks of the ships would have been the surgeon's quarters and he would have been given the tools of the day to fix or heal his patients. This would have been very crude compared with today. The environment was harsh. There was only one surgeon's mate, if any assigned to each vessel. A trip to the museum of the USS Constitution in Boston and USS Constellation in Baltimore speaks volumes on the conditions below decks in the Revolutionary navy. Very unsanitary indeed. The secretary of Navy was directly involved in assignments of commissioned officers to the vessels and hospitals on land. There was no hospital ships at that time. Early sick bays were not any more than a cutting table in a room away from the gun batteries below deck. There is a saying in the navy medical department that "…those were the days when the ships were built of wood and the men were built of steel". A literature search found very little on the subject of revolutionary war navy medicine. There seems to be little on the subject even more recently. The location of the naval academy in Maryland was convenient for navy medicine to learn from the major universities in Baltimore whom specialize in medicine before the Uniformed School of Health Sciences was created by congress in Bethesda. Officers whom wanted to become medical officers could go to school here in Maryland after becoming familiar with the surroundings at the naval academy. Only the top graded graduates would have been offered these assignments.

7. Barbary Wars

   The Barbary wars were conducted by the United States against Tripoli in

the Mediterranean Sea from 1799-1804. During this time the United States

Navy Medical corps was defined by every ship having a surgeon and

surgeon's mate. The required term of service was 1 year and there was a lot

of turn over. Only one surgeon remained in the Naval service as a career.

Surgeons could of course make a more lucrative career in private practice.

Also there we more hopeful cases to attend to in the private sector. The Navy

was hard on men and many diseases were direct causes of the environment

on ship board and in foreign ports of call. The sick bay of early ships was not

very well developed with usually just a table for surgery below the gun deck

with minimal lighting. Life on ship was cold and wet causing much disease.

There were many accidents on ship which accounted for deaths of the

crewmen. And then there were deaths from direct conflict with the enemy.

The Navy lobbied successfully to have a shore hospital to tend to the men

after sea service. The funds for naval hospitals were taken out of their

paychecks at 20 cents a month since they were the direct beneficiaries after

and during sea service. The captains of the ships relied heavily on the

surgeons they chose to keep the men in good health. The surgeon was a

source of great hope for the men once they entered combat. A healthy

surgeon could save many lives and they rarely were engaged in the warfare with the men. There is a story of one surgeon who requested to be assigned with the men who were taking over a frigate in Tripoli. He wanted to be there with the men to keep attend to them immediately. His captain reluctantly granted him permission, but he was the first surgeon to go into combat operations on ship. Usually when ships engaged in gunfire the surgeons were below deck attending to men. This man would have been the forerunner of the Navy corpsman who served with the marines in shore assaults in World War II and Vietnam. Some of the ships commissioned in the Navy during that timeframe were the United States, Constellation, Adams, President, and Constitution. Many ship captured from foreign nations such as France and Tripoli were converted to other uses and the first hospital ship was among these captured ships.

8. War of 1812

The War of 1812 was fought between the British and Americans on American soil. Navy medicine was available on every ship by this time and the assignments were done by the secretary of the navy under James Madison. Many engagements took place requiring the medical officers treatment of patients. They kept logs on their patients with names, ranks, and treatments, remedies, and dispositions for pensions related to health or injuries. The medical officers also kept the body counts during various engagements and reported them to the secretary of navy whom in turn contacted families. The life during the war of 1812 of a medical officer was determined by his skill level and the amount of time he could endure the conditions. Not all surgeons stayed with the fleet. Some went into private practice. Many tried to obtain higher ranks after gaining experience on ship. A few played politics to obtain the sweetest assignments by gaining favor with the secretary of navy at the time. The major engagements during the war of 1812 were the Battle of Bladensburg, Battle of North Point, Bombardment of Fort McHenry, and Battle of New Orleans all of which had navy vessels with medical men aboard. Sometimes they had to come ashore to tend to men in the Army like at Bladensburg. They went where they were ordered and needed the most. At New Orleans they tended to the few wounded of General Jackson's army and many navy men fought along side Jackson when the lone American vessel bombarded the British from the sea. The wounds were mainly from grapeshot and cannon fire. Amputations had to be done at times for many of the less than 15 soldiers Jackson lost. The British lost over 2000 in a few hours time due to poor planning and Jackson's quick thinking to trap them as they advanced in the open and his troop remained under cover in a makeshift trench. After this battle, the British were through in America and the war was over.

9. Civil War

The civil war was a very costly war in terms of human life. During the civil war more than 23,000 men on both sides were killed at Antietam battlefield in a single day. The technology of war increased with rifled barrels, Gatling guns, rifled cannons, and improved ranges. Medical technology also greatly improved during this war. The ambulance was born to ferry patients off the battlefield to save their lives. The ships were ironclad to protect from side shot cannons. Navy men still had the diseases that were normal for the day, but they could get treatment faster because steam driven ships were faster than ships under sail. The doctors on the battlefields had better medical tools to perform surgery but it was still crude conditions for sterilization. Nurses like Clara Barton were helping tend to the wounded on both sides. It was clear that the prison camps lost many soldiers to disease and lack of proper medical attention. By the wars end, more men died in the civil war than any of the previous wars America had fought in. It was very costly in human terms and this was disconcerting to President Lincoln a devout Christian. His speech at Gettysburg painstakingly described the holy ground where they shall not have died in vein. He was very upset at the cost of the war. He personally wrote a letter to one woman who lost five sons in the civil war and thanked her for her sacrifice for freedom. He paid for his support of the war with his own life and medical doctors could not help him from his mortal wounds in Ford's Theater by disgruntled southern actor John Wilkes Booth.

Booth was later shot to death in a barn when he was captured with his

conspirators. His identity was confirmed by the doctors on the scene.

Many of the veterans of the civil war lived well into the 20th century and the

wounded are memorialized at the many monuments to the civil war all over

the eastern part of the country on battle sites. Had medical practices been

more crude without improvements many of these men would not have

survived.

10. Spanish American War

The Spanish American war was a very quick war started with the explosion the USS Maine in Havana harbor. There has been discussion that the explosion may have been caused by a boiler mishap rather than a well placed bomb on the hull. Navy medicine was there and supported the Rough Riders battle up San Juan Hill in Puerto Rico. The war did not last that long that we received too many casualties. The length of war is proportional to the loss of life. The longer the conflict the more life is lost. Thank God this war was very short. There are various books on the subject by Theodore Roosevelt who led the Rough Riders. Today, Roosevelt Roads naval station is on the northeast tip of Puerto Rico and we maintain a leadership presence there.

11. World War I

World War I was basically a trench war. The huge Dreadnaughts on the ocean rarely fought. Naval battles were mainly between the British Royal Navy, US Navy and the German Navy. Germany used U-Boats for the first time to terrorize the shipping trade. The ships had sick bays by this time. The first of the naval hospital ships were commissioned during this time frame to support sea medical operations. An old tanker ship was reconfigured into a hospital ship. Medicine, in general, had made some advances but major vaccines were yet to be produced for many diseases. The life of a sailor was still fraught with disease and discomfort during sea duty. The sea itself posed certain medical risks that may never go away due to the cold, dank nature of salt water and it's affect on human living organisms. The Germans had huge battleships in it's fleet and the British and Americans countered by building as many as they could. The Dreadnought fleets of the world were status quo. Each ship was outfitted with a medical sickbay inside the ship where the ships doctor could work his miracles. Shore hospitals were also developed better during the first world war and congress controlled them and the appointments of officers through BUMED as they always have. Ships as small as minesweepers had medical sickbays on them too. The men got sick from the standard sea diseases of the tropics and water borne environment. Woodrow Wilson as a university president would have know about medical advances in educating medical schools and would have ensured that these were used in the fleet.

12. World War II

Much of world war two at sea was spent in either the North Atlantic or the Pacific. The marines had navy corpsmen on the beach landings where the army always had it's own medical units and trained individual soldiers. The ships were now evolved into living cities and small towns. Today's ships are sometimes named after cities and towns. Medical treatments advanced during the 1920's and 1930's before the war. These treatments were used on the ships. The movie "Mr. Roberts" shows how a ship's doctor interacts with other crew members on a Liberty Ship in World War II in the Pacific. This movie showed a cruel harsh captain and how the medical officer and crew coped with his over-aggressive style of leadership. In the climax scene, Ensign Pulver reluctantly saves the mean captain by taking out his appendix on a remote island talking to Mr. Roberts over the communications system.

The Marine landings on the Beaches were the most horrific scenes of World War II and they took the most casualties. Allied military victories at Saipan, Guadalcanal, Okinawa all came at very high prices in human life on both sides. The Japanese were formidable foes and did not give up easily. What they lacked in size they maintained in spirit and durability. The navy corpsmen were right there with the men on these landings. Sometimes the Army would make a landing and they would be supported by medical units in the army and navy. The Mercy played a role in this but I am not sure exactly where they would have been stationed certainly not too near the battles in theater part of the war when it was clear that Japanese submarines would attack any kind of shipping – hospital ships or POW ships. Total tonnage sunk was what they cared about, not international laws of restrictive nature on sinking hospital ships. During sea battles in the pacific there often was no law of the sea. The Germans frequently showed kindness in battle with truces during holidays like Christmas because we shared Christianity with them (outside of Hitler's brand of Aryan Christianity).

Reality was far from the Hollywood version of the story the medical staff and nurses in the Pacific theater had to endure. The Japanese considered Americans barbarians and the ruthlessly killed nurses and doctors in hideous fashion when they captured them during the early parts of the war in the Pacific. Some nurses were decapitated and their bodies (breasts and male sexual organs) were completed removed by Samurai sword. There were also methods of torture that included a glass pipe inserted into the male captive's penis and broken into pieces. This would be especially horrific to a doctor who knew the potential for damage of this sort to the sexual organs. These gruesome stories only served to infuse the Allied troops to fight harder against Imperial Japan's ruthless wartime behavior and treatment of POWs. The Asian mindset of tremendous torture methods was continued in following wars with North Korea and North Vietnam whom were manned by Chinese regular Army under communism. The regard for life under these regimes was barely what human rights are today in America. From the Japanese perspective, they consider us barbarians because of the tales of the Vikings, Scotsmen, Europeans, and Americans in past wars. We must remember they consider themselves a single race of people untouched by the outside world except Japan. Our methods of gunpowder did not reach them until the 1800's. They were dedicated Samurai warriors for many years with honor. Honor is very important to the Japanese culture. Karate is a self defense mechanism only not to be used for offense. Japan also tries to purify it's young by shipping out most of the mixed race children. The Pearl Buck Society helps these children find their real parents in a mission of mercy. Chinese Buddhism once influenced Okinawa and it is visible in the culture, architecture, and religion. The Japanese are no longer allowed to practice state run Shinto religion which led to World War II. Interesting enough, they call this the period of American occupation and believe that in their thousands of years history the Americans will fall from grace and leave their land in the future at some point restoring Japanese greatness and pride in the world. They are very proud of their technology and western ways they have acquired from the US and free democracies of the world.

The Germans in World War II were just as ruthless as the Japanese if not more. Dr. Joseph Mengela experimented with Jewish prisoners before they executed them. America did not of this type of mutilation of our POWs. In fact it was said for the best medical treatment and POW conditions it was best to be captured by the Americans. Italian prisoners of war were held in Arizona and treated with kindness. America did not want to get into this war, but we sure

finished it as a winner.

One of the bloodiest days of the European theater was on D-Day and my

grandfather was there with the 51st Highlanders. The naval armada off the coast

of Normandy in Operation Overlord was the largest ever in mankind. It was June

6, 1944, the greatest day in history of battle in the 20th century. The naval

hospital ships were stationed in the English Channel and soldiers were treated

on the battlefields and then moved to the rear formations, if they survived. They

Would have been on the hospital ships for the most critical surgery and then sent

to hospitals in England, Scotland, or the States for recuperation and return to

duty or deactivation back into the public. The highest casualties were on Omaha

beach for the Americans that day due to 27 of the 29 tank armor support sinking

in the waves off of Omaha Beach. They were massacred by the German gunfire

from the bunkers which were missed by air bombardments. It was the US Air

Force who took the most casualties in World War II over the skies of Germany.

13. Korean War

Much of what the public knows about the Korean War medical units is perpetuated by the TV series MASH – a humorous approach to telling the story of an Army Mobile Medical Unit. Naval hospital ship units were there and worked in unison with all medical units (Marines, Army, Air Force) in evacuation mode. The surgery that could be performed on the hospital ship was done in a more stable and controlled environment for the patients and they probably received better treatments after initial triage and surgery in the field units on the peninsula of Korea. The navy did not lose as many of it's people in Korea because it was basically a ground war and the US dominated the seas. The Air Force had casualties in the Air Campaign against Russian MIG 15 fighter aircraft and we had the capability to Airlift using the C9 Nightingale to Air Force hospitals stateside. The Naval ship platforms were very safe and stable because they had so many surgery rooms and all the equipment in an environment where there was no threat to the staff or crew members. They were truly on a mission of mercy. Patients could be evacuated after immediate surgery to Yokuska, Japan, Guam, Anderson AFB, or Pearl Harbor. Tripler Army hospital, Hickham Hospital, and Peral Harbor Naval Hospital were all available on Hawaii's Oahu island. This would have been a better place for recuperation from injuries since the patient could either fly back to the states or be sent back to combat if well enough and able.

In the series MASH, Japan is usually the place the troops are sent for rest and recreation. It was also the location of a large military hospital complex by the three services. The Marines always are welcomed at the Navy Hospitals the same way they train at the Naval Academy as a tradition, but they have their own OCS commissioning program and University at Quantico Marines Corps Base since those days.

14. Vietnam War

What I know about the Vietnam War and Navy Medical comes primarily from a previous supervisor who was a Navy Corpsman attached to the Marine Corps and what I have heard from Colonel Oliver North on the matter of how much love the men had for their corpsmen. The navy corpsmen were often unarmed and protected their patients on the battlefield. I always got the feeling that these guys were real heroes every time they saved another man from death which was quite often. My supervisor never directly talked about Vietnam except that he was there and it was hot at times. He spent time on Japan at Yokosuka Naval Base and we have a large MTF there for our servicemen. He was one of the nicest people I ever worked for and he was a very hard driving but fair man. He was very dedicated to the mission of Navy Medical and this came from his duty in Vietnam when he was young. He was open minded and a very good leader. He always gave me hope that I would someday lead my own database division. I just felt like he valued life and all that comes with it. He stated that several times he was on the USNS Comfort hospital ship. I never questioned his ability to lead us and learned as much as I could from him about mid-level management at Navy Medical. He opened doors for me because of my loyalty to him and Navy Medical.

The Navy also dominated the seas in Vietnam with the exception of Russian submarines and Chinese submarines whom now have entered the nuclear submarines race. Usually hospital ships are off the enemy's target list, except for the Japanese. The large cross on the side of the ship designates that they are carrying wounded. The South Vietnamese navy was not able to provide for a hospital ship so the Mercy saw action in this war too.

Navy Medicine was there when the Apollo 11 astronauts returned from the moon landing and they quarantined the astronauts on a carrier at sea in 1969. President Nixon met them there and shook their hands.

Navy Medicine also took care of the POW's after Vietnam in 1974-5 from Hanoi when they arrived at Andrews AFB. Senator John McCain outlines this in his book Faith of My Fathers.

15. Presidential Patients

One of the most famous cases in Bethesda Naval Medical history is
when President John F. Kennedy was taken to Bethesda Naval Hospital for a
post-mortem after he was assassinated in 1963. There have been several
books and movies on the subject. President Kennedy was examined by
Naval doctors there after he was flown back from Dallas, Texas. Every
President gets his choice of military hospitals and since JFK was a navy man
(PT 109 which I built at 8 years old) he chose Bethesda, naturally. There will
never be another JFK as president in my mind as he endeared the country,
suffered greatly from his own hidden infirmities, and died a tragic death in
office. His book Profiles in Courage[3] is a must read by everyone in
government service. He wrote it about public persons who sacrificed greatly
for the common good of the nation. He also wrote it on his sick bed when he
was in the hospital.

It has been suggested that there was a cover-up at the scene of the
autopsy. I believe it is very hard to cover up medical evidence and against
the nature of the military to cover up anything that important to the nation.
The Warren Commission Report and other books tell what happened that
day. There was confusion and a lack of chain of command, but I think this
was natural after a president had been slain. It was one of the saddest days

---

[3] Kennedy, John Fitzgerald, Profiles in Courage, Random House, 1962.

of my childhood life and it was the same day that I learned about death at a tender young age of 7 years old in the first grade in Topeka, Kansas. My mother cried like a baby and told me that president Kennedy had been shot. Mom was not a Catholic but she really liked president Kennedy since our family were Democrats and he was very much for civil rights and rights of the underprivileged. My father was overseas in Japan at the time for the Air Force. I could not make sense of the whole situation but cried just like my mother did for our slain president. Little did I know that many years later I would work at Bethesda Naval Medical Center for 10 years.

Other Presidents who were treated at Navy Medical were Jimmy Carter a navy man (submariner), Ronald Reagan (military actor), George HW Bush (navy flyer in WWII) in recent times. Eisenhower attended the US Military Academy and used Walter Reed Army Medical Center (WRAMC) as his hospital of choice.

Once while I worked at Navy Medical Data Services Center, Ronald Reagan arrived at Bethesda by helicopter and the staff and people rushed to see him and shake hands as he got off the chopper. He often looked out the windows and waved at folks with his friendly way. He was a real man of the people until he was shot in 1982 when he was rushed to George Washington Medical Center in Washington DC (it was closer than Bethesda that day and out of the normal routine for security purposes).

I always thought it funny that the most powerful man on earth had to submit his authority and trust to the medical doctors for that time that he visited them. The fate of the entire free world was in the hands of the doctors doing surgery on the presidents. Most doctors would say that they have more pressure saving a child and that every patient is equally valuable. Is that enough pressure for anyone! The Apostle Luke, a medical doctor in his own right, would be proud of them.

16. The Presidents at NNMC

Every president has the privilege of being served by the people at one of the Washington DC Area military hospitals. Some chose Walter Reed Army Medical Center, some chose Bethesda Naval Hospital. President Reagan went to George Washington University Hospital when he was shot by John Hinckly in 1982. The Presidents usually helicopter in to the Naval Hospital on Marines 1 for the appointments with their doctors and all the staff perks up to treat the First Patient. He is greeted by young and old alike who want to see him in person and hopefully shake his hand. The NNMC stands for National Naval Medical Center and it is really a special place. During Vietnam soldiers were sent to the NNMC to be treated after coming home through Dover Air Base or some other MAC (Military Airlift Command) transport base. They may have been treated on the USNS Comfort or the Mercy at sea for initial wounds and transferred to Bethesda. Unfortunately, only the most critical worst cases were sent Bethesda. Bethesda also served active duty military and dependents stateside. The location across the street from NIH was very strategic in that the research done at NIH could directly benefit NNMC and Navy Patients. People form all over the world come to NIH for research cures for various diseases and infections. NIH is world known in the medical field and the NLM (National Library of Medicine) houses the world's largest medical library. NNMC is a military post like most bases and houses the

Military School of Health Sciences which is the graduate school of military medicine for all the services. The old tower building was built in the 1940's and parts are still operational today although engineers have been trying to condemn it as unsafe for years. The mall building is new and was built in the late 1970's and early 1980's. Every medical services organizational department is located inside a shopping mall type of building which was new for most hospital builders at that time. Pharmacy, radiology, urology, gynecology, kinesiology, physical therapy, nuclear medicine, obstetrics, and many other clinical departments are located in a mall where patients have plenty of room to sit in the large waiting areas in the center of the malls. Interesting enough, all the computer systems are build to reflect the clinical organizational structures that they support. There are grand escalators and staircases inside the mall and two decks. Walkways and paths to the clinics make the tour a pleasant one. The helicopter pad for critical patients who are transported by air (medevac) is located next to the main administration building on the campus.

Also on the campus are what is called tenant commands. A tenant command on any military base is a command that rents that space when the primary mission belongs to another command. NMRI (Naval Medical Research Institute) is one of the tenant commands at Bethesda and provides the latest in medical research for the US Navy. I always assumed NMRI would work closely with NIH on many developments in lab experiments they would conduct. We had several seminars in the NMRI auditorium as they we well equipped to educate the staff and public. Another tenant command was NAV SUP or Naval Supply Services. They gave us our medical supplies and the data center inventoried them. NMDSC was the Naval Medical Data Services Center another tenant command. They provide Navywide data services to all MTF's. Special Services was tenant command that kept the staff in good fitness. They had a gymnasium and recreation services to built teamwork and unity of each command on base. We (NMDSC) competed in many softball and basketball tournaments with yours truly as the coach or co-coach whatever it took.

Parking at NNMC was always a headache as one of the largest hospitals in the world. At one time parking was free but now it is a sticker and pay as you go basis with free parking for visitors. Staff parks in the expanded new parking lots. Metro has a stop at the NIH campus, but that is a 10 minute walk from NNMC.

17. Congressional Appropriations

Congressional Appropriations for the entire navy are funded by the House of

Representatives and Senate as a normal appropriations bill in the Armed

Services Committee in the spring of the preceding year to which the money is to

be used. The federal fiscal year is from October to October. The State Fiscal

year is from July to July before the Federal fiscal year offset by 7 six months

Sometimes bills and appropriations are held up before the October Fiscal year

change. When the government is said to have run out of money this means the

bills are not signed by the Oct 1st deadline for the current fiscal year and have to

be continued into the new fiscal year. This is called a continuance. Navy

Medical receives its apportionments at BUMED due to the hard planning division

work of the senior staff who work with the hill in determining their future out years

budget needs. The money received by BUMED managers is funneled into the

various units to pay for personnel, transportation (Military Sealift Command),

materials, supplies, buildings, and other assets. There are many computer

systems to keep auditing track and accounting of these expenses using standard

accounting practices. Some ways to track the appropriations process are

through the Records Office at the Congress where you can obtain a copy of the

bill, through the Congressional Quarterly who writes on operations. through the

Government Operations committee members and chairperson, and on CSPAN

Cable TV. Usually the Senate proceeding and special sessions are not televised. Agencies in Washington and congressmen watch CSPAN themselves to keep up on votes and the entire process. GSA is the agency who manages government telecommunications acquisitions and FCC manages the telecommunications system. OMB manages the funding of these systems.

The best citizen's watchdog on congressional appropriations is CSPAN live and then the documents office at Congress which will give you a copy of any bill you request over the telephone. House bills and Senate bills and hearings are given out separately and some documents are sold to the public through the Government Printing Office and their website www.gpoaccess.gov. GPO is the printing arm of Congress. They have a good book selection also on government operations. GPO is located on North Capital Street in Washington DC and listed in the telephone directory under government services. They also produce and distribute CDROMs to the public for various published efforts and have an internal CDROM division who will publish documents and books on software for the government agencies in Washington. I managed one of these products during my tenure in Washington which became the best selling CDROM in GPO history at that time in 1995. It was a regulatory CDROM called the GSA FAR/FIRMR CDROM by my agency GSA.

The Congressional Quarterly magazine and National Review also are good resources for appropriations issues. OMB has a website that is useful in all money matters for the federal government. Realize that Congress has a different budget office separate from OMB and OMB works for the president not congress which is why numbers vary from year to year between the different branches of government. The president issues a proposed OMB budget every January and congress works with it in their budget proposals. The numbers are usually reconciled over time to come to an agreement before voting takes place on the amounts.

Defense appropriations take priority during wartime and all appropriations for all federal agencies are usually done a year ahead of time. Some major government acquisitions are programmed into the budget for multiple years and are known as multi-year acquisitions and appropriations. These usually affect the local level states and counties which employ the people building the products being acquired. This is also known as pork projects when some states have a larger share of projects than others. The Senators and congressmen compete for the resources in the appropriations for their state's constituents. Some lobbyists try to influence the process with large donations to campaigns where the senator or congressman vote will help that lobby.

18. BUMED Organizational Structure

BUMED means Bureau of Medicine and Surgery. In more recent years it
is synonymous with Naval Medical Command. It specifically refers to the
headquarters location at 23rd & E Streets Northwest , Washington DC.
It has been located there since the civil war just across the Potomac River
from the Pentagon a scant 10 minutes away and within 10 blocks of the
Whitehouse, monuments, and OPM. The BUMED organization structure has
been published on a website and is published here from that website. It is
organized by Codes as are many government installations. This way when
someone says Code 041 it has a specific meaning. It really means internal
mail code for distributing mail to various offices.

Figure 1. Logo from Sea Power 21 webpage

The BUMED organization structure has evolved over the years with every organization change but is now pretty well set as shown in figure x. We served as the Data Processing Center for all the codes and worked for every one of them by implementing their corporate data processing infrastructure and resource management needs. Our data reflected the structure of the headquarters organization at BUMED and we knew all the people there who helped us build our databases and systems for them.

Figure 2. BUMED Internal Organization Structure

Pastoral Care (M00G)

Staff Judge Advocate (M00J)

Medical Inspector General (M00IG)

Professional Integrity and Medical Ethics (M00RIE

Navy Dental Corps (M09B-DC)

Medical Corps (M09B-MC)

Navy Medical Service Corps (M09B-MSC)

Force Master Chief (M00FMC/M09-HC)

Navy Nurse Corps (M09B-NC)

Research and Development (M2)

Fleet Operations Support (M3F)

Medical Operations Support (M3M)

Logistics (M4)

Information Management (M6)

Education and Training (M7)

Resource Management/Comptroller (M8)

Environmental Health (M11)

Worldwide Navy Medicine Facilities

Reserve Affairs (M10)

Office of Knowledge Management (M09BK)

Relationship to HHS Public Health Service

Navy Medicine is a sister organization to the Public Health Service.
Surely, the uniforms look very similar. The head of the public health service
is the National Surgeon General. He is the president's advisor on all national
health care issues. The Public Health Service has office's in Rockville near
Naval Hospital Bethesda. The two organization's share research and
information but otherwise are two separate functioning organizations. The
Center for Disease Control in Atlanta is under the Public Health Service not
the military.

Relationship to NIH Research and Development

Naval Hospital Bethesda is directly across from NIH on Wisconsin Avenue and shares some research with NIH. They also share computer facility resources and other resources when needed as most government agencies do. NIH medical research is the best in the world. The National Library of Medicine (NLM) is a resource that both share. The Uniformed School of Health Sciences also shares the NLM. The NLM is open to the public and can be used by anyone with a valid research effort. Physically NLM is located on the NIH campus but there is computer access from anywhere in the world.

Relationship to Health Insurance Companies

There is no relationship between Navy Medicine and Health Insurance companies as Navy Medicine is not chargeable. The only people allowed to utilize Navy Medicine are the military and political appointees in the federal government. The Navy probably likes it this way as they really do take care of their own. More and more the Navy contracts out certain functions to private contractor companies who can help with the work load which includes medical care.

Relationship to Pharmacies

Navy Medical has it's own pharmacies in the Naval Hospitals and Clinics in the command structure and they can dispense medications and drugs with the best of them. The costs is inexpensive compared to private doctors. Any dependent or military person can receive legitimate prescriptions from the Navy Pharmacy when they need it. The patient records are kept for 5 years and then shipped to St. Louis.

19. US Surgeon Generals

The United States has many surgeon generals. There are the surgeon generals of the military and the Public Health Service. The latter cover all of the United States Citizens on health issues and reports directly to the president. Surgeon generals of the PHS are listed in figure x and have been in office since 1987.

Figure 3. Surgeons General of the Public Health Service

John M. Woodworth	March 29, 1871 to March 14, 1879
John B. Hamilton	April 3, 1879 to May 31. 1891
Walter Wyman	June 1, 1891 to Nov. 21, 1911
Rupert Blue	Jan. 13, 1912 to March 1, 1920
Hugh Smith Cumming	March 3, 1920 to Jan. 31, 1936
Thomas Parran	April 6, 1936 to April 5, 1948
Leonard A. Scheele	April 6, 1948 to August 2, 1956
Leroy E. Burney	August 8, 1956 to Jan. 29, 1961
Luther L. Terry	March 24, 1961 to Oct. 1, 1965
William H. Stewart	Oct. 2, 1965 to August 1, 1969
Jesse L. Steinfeld	Dec. 18, 1969 to Jan. 30, 1973
(Acting) S. Paul Erlich 1973 to 1977	

Julius B. Richmond*   July 13, 1977 to May 1, 1981

C. Everett Koop*   Jan. 21, 1982 to Sept. 30, 1989

Antonia C. Novello   March 9, 1990 to June 30, 1993

M. Joycelyn Elders*   Sept. 8, 1993 to December 31, 1994

(Acting) Audrey F. Manley   Jan. 1, 1995 to June, 30, 1997

(Acting) J. Jarret Clinton   1997 to 1998

David Satcher*   February 13, 1998 to February 13, 2002

Richard Carmona*   August 5, 2002 to the present

*Surgeons General marked with an asterisk were not members of the

Commissioned Corp at the time of their appointment.

20. Commanders

Commanders of Navy Medical

The commander of navy medicine is usually also the surgeon general of the navy. This person is usually an admiral in rank and has earned his position through a career of successively more difficult position sin the navy with great achievement and honor. He resides in Washington DC at the Naval Observatory and works at the 23rd and E street Navy Medical Command Headquarters building. He is an advisor to the joint chiefs of staff at the pentagon which is located across the Potomac River from BUMED on all navy medical matters. It is a ten minute car ride through Washington. Telecom is available for direct communications from him to the Pentagon.

Structure of Command

The structure of Navy Medical is as follows in general. Some of the names have changed since the structure followed this hierarchy. The PHS cover all the civilian world in the United States.

Figure 4. US Navy Medical Organization Structure

President -------------------------------- PHS Surgeon General

Congress

Surgeon General of Navy

BUMED

Naval Hospitals

Naval Clinics

USNS Mercy

USNS Hope

Navy Units

NMDSC

NMRI

Tenant Commands

Admiral Lewis Angelo and the perfect rating

I received my only perfect rating in government service under
Captain Lewis Angelo and Al Stonebraker.   I worked harder for them than
anyone I knew before. They rewarded my loyalty with an outstanding
rating in the civil service. The interesting thing was that I was in a position
that was newly created by OPM. The rating was something I a have

always been proud of and I am thankful that they appreciated me as they did. It was the only time in my career that I worked for an Italian commanding officer and he was very dynamic and good with people. Everyone admired him greatly for his enthusiasm and livelihood. I remember that when he gave awards he had the people gather around in the hallway once a month and gave his awards then in front of everyone in a semi formal setting where he stated how proud he was of the employees. I have never seen this type of leadership since then. Captain Angelo became a Commodore and Admiral as I heard before he retired. He sure inspired me to perform well.

21. The Combat Navy Corpsmen

Working for a Navy corpsman is a unique experience. They are deployed with the Marines in the field at the hot spots of the world and go into battle to save other men's lives with knowledge of medicine and little else to protect them. They are a Marines best friend when wounded. They protect the wounded and fight if necessary to protect that wounded man. A good combat medic is worth his weight in gold to his comrades. He has the battlefield awareness to both take life and give it under great duress. His failure means one of his own men dies. They seemed to be extremely loyal to other corpsmen who were their friends. This goes back to any time spent in combat with people you know. Vietnam was no different than any other war except that the hospital ships were the transfer point to homebound soldiers who were wounded. Army hospitals did the bulk of work on land and transferred critical patients to the ships for transport.

I learned more about good management from this one guy than all my other bosses I have ever had in government or the private sector. He used to always say "Economics are the key to understanding the world". He sent me to graduate management school and gave me a real lesson in Christian leadership. He expected a lot from me and I delivered because I did not want to let him down. He enjoyed teaching me how to build "bulletproof" computer systems as he called them. I taught him about James Martin and had him reading his management books; he was the British ADP guru I learned about at University of Maryland in the IFSM department. The Navy graciously paid for this for me in return for my civilian services. He did not talk much about his combat corpsman experiences except that he was stationed on the USS Okinawa. At the time I knew him, he was in the MSC (Medical Service Corps) and naval reserve. He respected enlisted men and women and knew how to maximize on their potentials. This was because he came up through the ranks himself before he was commissioned as a Naval Officer. He was always for more education and using his people wisely in their best capacity. He rewarded hard work with cash bonuses and promotions. He also kept learning the next level job he needed to know to make himself more valuable to the CO. He became the executive officer from a division officer with our help. When he retired he became a CEO in the research and development company he went to work for.

Today, I emulate his hard charging management style and salute him as a good friend of mine whom I send Christmas cards every year. I realize as a manager he was just doing his job, but he did it so well he left a lifelong impression on me. He guided my baptism under fire through the new positions I held as database systems engineer and he was very trustworthy in my book. I earned an outstanding rating and cash bonus as well as several letters of appreciation and employee of the week from him. Thank you LCDR

Alan F. Stonebreaker.   You made me a better person, supervisor, father, and husband. You were one of the many big brothers and father figures I had during my tenure at Navy Medical. You are the best in my book.

22. Field Surgeons

Field surgery goes back to many previous wars. The most improvements in field surgery came during the civil war when technology advanced due to the sheer number of casualties. The field horse drawn ambulance was created. Surgeons used better equipment because manufacturing was improved. Led shot made wounds terribly large and patients suffered extremely greater than with other weapons of previous wars. The naval ships had doctors on board to handle most medical situations as in previous wars but the number of battlefield casualties exceeded even President Lincoln's imagination during this conflict. People actually went out to the battlefields in Northern Virginia to watch the war take place like it was a Broadway show and were then revolted to see such bloodshed. The grapeshot, mini-balls, and cannonballs were all devastating to human flesh and bones. The surgeons worked wonders to save as many men as they could through amputations and other extreme surgical procedures. The first triages were used in the civil war to determine who should be treated first on a priority list. There were films used to document the wounds since the camera was a new device on the battlefield. Often uniforms cause infections when a bullet entered a man. Some of the cannon weapons were now rifled which made them even more effective in combat at destroying men. A row of men could be cut down by one rifled bullet shot from the side from the new rapid fire carbine guns the Northern Army of the Potomac was using. This tactic was used when men marched in multiple filed rows on the battlefield and were outflanked by the enemy. The South had the older single shot muskets since the north had the factories to produce better weapons. Field surgery was a fast and furious occupation in the civil war and we learned a lot from it as a nation. The wars of the 20th century were also brutal and field surgery was critical to get patients to the recovery phase of the wound. World War I saw the use of gases and trench warfare. Treatments for all types of diseases related to the environment of mud and water had to be developed as well as for the typical bullet wounds. The Navies had fights at sea but now ships were made of metal and wounds were far greater in symptoms and damage due to the damage of the metal parts exploding into sailors. The Navy surgeon always had to deal with water borne diseases. He now had specific areas of the ships called sick bays where he could treat his patients with the latest technologies and remedies. He had an assistant who learned from him and moved up in ranks when requirements were met. The film Mr. Roberts shows the interactions of a World War II Liberty ship doctor and crew very well. The ship's doctor was always well liked and usually a friend of all the sailors for his healing abilities and nature.

23. Aerospace Flight Surgeons

It was quite well known that one could earn several special pays by being a
flight surgeon in the military. There is a special pay for being an air crewman and
a special pay for surgeons. Then if that person was in a combat zone another
special pay would kick in extra cash for the home front. Flight surgeon's usually
are concerned with aero medical concerns of pilots and crews. They also have
responsibilities for aero-medical evacuations. During Vietnam medical
evacuations were performed on Bell Huey helicopters and soldiers were ported
to a field hospital or ship board hospital for transfer back to a hospital in a nearby
friendly port in Japan. The C9 Nightengale was the name of the aircraft used for
aero-medical evacuations by the Air Force specially equipped with bunks and
medical equipment. The best surgeons usually become this type of surgeon and
advise both the fleet and others on air operations health and welfare. The
education must include both medical school and flight education and physiology
of flight. The doctors would not be expected to fly the aircraft as pilot in the role
of flight surgeon, but serve aboard an airplane or aircraft carrier. Flight surgeons
certify the health of the pilots on ships and ensure they are flight ready. They
also have the power and authority to ground any pilots who are not fit for air flight
duties on any given day or night mission based on medical or psychological
reasons. The modern flight surgeon can get trained at one of the Air Force
facilities or others in Texas after graduate medical school and officer candidate

school. This is a remarkably good find for the services as they may not have to pay for the medical school training. If they do pay for the training it may be through the Uniform School of Health Sciences at Bethesda, Maryland. This school has the resources of NIH, The National Library of Medicine, and the Naval Hospital Bethesda at it's calling. Students may use any of these facilities in their education process. NAVAIR is the navy equivalent of the Air Force and flight surgeons will see duty with this group of fine airmen.

24. Typical Fleet Operations

   The Naval Hospitals and Clinics support the fleet from the shore for the

most part except during combat operations where the Mercy or Comfort will

be stationed off the coast of the combat littoral. The Mercy supports the

Pacific fleet in San Diego and the Comfort supports the Atlantic fleet in

Norfolk. Every ship has a sickbay within the ship where the ship's doctor

attends to his patients. This was chronicled in the World War II navy movie

"Mr. Roberts" where Walter Matthau played the Doctor on a US Liberty ship in

the Pacific and Ensign Pulver saved the ships Captain (Burle Ives) with minor

surgery on a benign appendix. Mr. Roberts was adored by his men for his

rational thinking and attention to their every medical needs. Two other recent

movies series "MASH" and "China Beach" describe recent day front line Army

hospital action in Korea and Vietnam, respectively. Both of these series had

some truths in them about front line surgical operations. It is quick and

sometimes it is dirty without the luxuries of the types of equipment found in

the 1000 bed hospital. The Mercy and the Comfort provide floating hospital

surgical platforms that simulate a 100 bed hospital. They are converted

tankers redesigned and commissioned in World War I to support our troops.

They're deployment is secret so as not to let the enemy know where

casualties may be expected. We never knew where they deployed during my

entire 10 years at Navy Medical. We may have had some staff members

who had served on them for short assignments but we were not a surgical support group only data processing. It's clear from the historical accounts of the Barbary Wars, War of 1812 that navy medicine has evolved from a small room below decks next to the gun ports with only a table and saw to a pair of large converted tanker ships. The crew of the current day hospital ships is from the MSC – Military Sealift Command and they basically drive the ship. The Surgical and Nurse crew is from Navy Medical and they do the heavy lifting of treating casualties on one of the many surgical rooms on board with the most modern equipment. The ships are outfitted with all the departments found in the Navy Hospital including pharmacy, radiology, surgery, recovery, nuclear medicine, aviation medicine, and any others. They support helicopter landings on the decks of the ships to bring patients to safety. Other ships in the fleet would have specialists also such as aircraft carriers would have aviation medicine specialists as well as general practitioners for all types of sea going diseases. When one tours some of the older ships that are moored around the United States as floating museums you can see the confines of the sick bay in that particular ship (USS Intrepid, USS North Carolina, USS Massachusetts, USS Constitution, USS Constellation, Ark, Dove). Submarines had a relatively small area designed for sick bay compared to the larger ships and some may not have had one at all in the older subs (US Torsk). Fortunately, we have become more humane in how we view combat casualties and our treatment of them has improved since the civil war. In Scotland, a tour of the Sir Francis Drake vessel in 1983 showed that the British ships of the time had very little space dedicated to the navy medicine function and it shared space on board.   Needless to say if you needed medical attention on some of these ships you may well have been in trouble because they all have limited resources and supplies they carry due to weight restrictions. The shipboard environment was also very harsh and certain diseases thrived in the locations and areas the ships were deployed to and were just more difficult because of the salt water environment the navy operates within. For many sailors, a government pension awaited them should they be discharged due to an incurable illness. Today, we have the resources to transport the sickest cases to the nearest shore facility and save that persons life before the case is leads to death. This was not always so before medical evacuation (medevac) helicopters.

25. Hospital Ships - USNS Mercy & USNS Comfort

The following excerpts on the two navy hospital ships were taken from the
Navy Medical webpage under the hospital ships[4] and are currently public
released information by BUMED and the navy. This information was <u>not</u>
classified and was found on internet. The author does not agree with the
lack of security on general information on these ships but it should be noted
that no detailed information is given in this public webpage write up from the
navy.

The two ships support the fleet operations in the Atlantic Fleet (Norfolk)
and the Pacific Fleet (San Diego). These ships are unarmed and designed to
hold a surgical command on each ship. They each have over 100 beds. This
is how hospital sizes are measured in the United States (bedsize). They are
literally floating hospital units that support operations anywhere in the world.
If an enemy knows where the hospital ships are headed, then they know
where we expect casualties. This is why their operations may be classified
with an entire naval operation.

"The first MERCY (AH 4) was built in 1907 as SARATOGA by William
Cramp & Sons, Philadelphia, PA. An Army troop transport in the first nine
months of World War I, she was renamed MERCY and converted to a

---

[4] www.mercy.navy.mil accessed on 8 July 2004.

hospital ship at New York Navy Yard, Brooklyn, N.Y. She was commissioned on 24 January 1918.

MERCY initially operated in the Chesapeake Bay, homeported in Yorktown, VA. She attended war wounded and transported them from ships to shore hospitals. On 3 November 1918, MERCY departed New York City, making four round trips to France, returning 1,977 casualties by March 1919.

For 15 years following World War I, MERCY served off the East Coast, homeported in Philadelphia. From December 1924 until September 1926, she was in reduced commission. MERCY was loaned to the Philadelphia branch of the Public Relief Administration in March 1934.

The second MERCY (AH 8) was a troop ship built by Consolidated Steel Corp., Wilmington, CA, beginning 4 February 1943. Launched on 25 March 1943, she was sponsored by LT(JG) Doris M. Yetter, NC, USN, a prisoner of war on Guam in 1941. MERCY was converted by Los Angeles Shipbuilding & Drydocking Co., San Pedro, CA. She was commissioned 7 August 1944, staffed by the Army's 214th Station Hospital.

Departing on 31 August, she arrived in the Leyte Gulf, Phillipines, on 25 October during the Battle for Leyte Gulf. Embarking 400 casualties, she transported the wounded to base hospitals in New Guinea. Over the next five months, MERCY made seven more voyages from Leyte to New Guinea, including transporting the Army's 3rd Field Hospital from New Guinea to the Philippines in January 1945.

In March 1945, MERCY reported for service in the Okinawa campaign. She and USS SOLACE (AH 5) arrived on 19 April at Hagushi Beach, Okinawa, embarking patients for four days despite frequent air raids and threat of Japanese Kamikazes. MERCY then transferred the wounded to Saipan, Marianas. She made two more voyages from Saipan to Okinawa the following month.

MERCY next made two voyages carrying wounded from Leyte and Manila to New Guinea. She reported to Manila in June for two months duty as station hospital ship. In August, she transported the Army's 227th Station Hospital to Korea as part of the occupation forces.

In October 1945, MERCY returned to San Pedro, CA. She transferred to the U.S. Army on 20 June 1946 for further service as a hospital ship.

MERCY received two battle stars for World War II service. She was struck from the Navy List on 25 September 1946.

The third MERCY (T-AH 19) was built as an oil tanker, SS WORTH, by National Steel and Shipbuilding Co., San Diego, in 1976. Starting in July 1984, she was renamed and converted to a hospital ship by the same company. Launched on 20 July 1985, USNS MERCY was commissioned 8 November 1986.

On 27 February 1987, MERCY began a training and humanitarian cruise to the Phillippines and the South Pacific. The staff included U.S. Navy, Army, and Air Force active duty and reserve personnel; U.S. Public Health service; medical providers from the Armed Forces of the Philippines; and MSC civilian mariners. Over 62,000 outpatients and almost 1,000 inpatients were treated at seven Philippine and seven South Pacific ports. MERCY returned to Oakland, CA, on 13 July 1987.

On 9 August 1990, MERCY was activated in support of Operation Desert Shield. Departing on 15 August, she arrived in the Arabian Gulf on 15 September. For the next six months, MERCY provided support to the multinational allied forces. She admitted 690 patients and performed almost 300 surgeries. After treating the 21 American and two Italian repatriated prisoners of war, she departed for home on 16 March 1991, arriving in Oakland on 23 April.

USNS MERCY, homeported in San Diego, CA, is currently in reduced operating status with a five day activation.

Figure 5. USNS Mercy

Figure 6. Medevac Copter on deck of USNS Mercy

The USNS Comfort[5] is home ported on the Atlantic coast and Atlantic Fleet. She is home ported in Baltimore and supports fleet operations around the world. This is convenient to the many public hospitals in the Maryland Baltimore / Washington metropolitan area including Naval Hospital Bethesda, USHS, Walter Reed Army Medical Center, and Malcolm Grow Medical Center (AF). There are many clinics and smaller MTF's at the military bases in the Maryland / Virginia area. There have been 3 USNS Comforts since 1907 in the Navy, each one serving the fleet before being decommissioned. The most recent one was commissioned in 1988. The first USNS Comfort was used as a WWI troop ship and then converted to use as a Hospital ship.

Figure 7. USNS Comfort

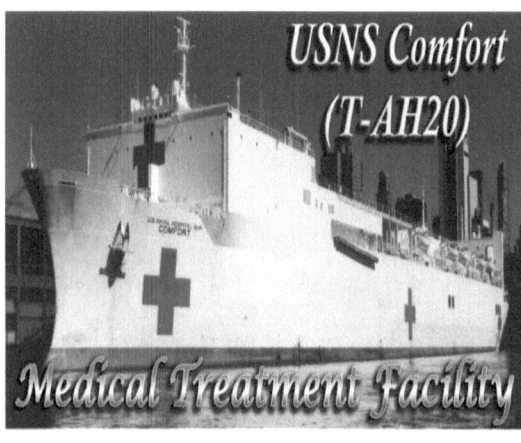

[5] www.globalsecurity.org/military/agency/navy/tah-20.htm accessed on 21 July 2004.

The USNS Comfort is a Mercy class ship being the second of that type developed by the manufacturer Bethlehem Steele. The crew complement is about 550 medical staff. The Comfort has been deployed in every major military operation since 1988 to the littoral seas near the action to take on the casualties of those conflicts including Desert Storm, Desert Shield, Sea Signal, Uphold Democracy, Noble Eagle, and Iraqi Freedom."

Figure 8. USNS Comfort Insignia

26. Medical Corps

The Medical Corps are the doctors who comprise the Navy Medical Command. Many of them are graduates from prestigious medical schools and many are board certified. We kept track of this in our database. As it turns out, the database is the perfect place to keep personnel information for management use. When a doctor got reassigned we always had the information for the assigning duty officer to use for that officer's background. We also kept designator information on the previous positions the doctor held in Navy Medicine and the times and dates he held them. We had rank, name, addresses, and many more fields on the doctors that could be used by management. We had their specialty and subspecialties, foreign languages, and college information. We had their PEB (Physical Evaluation Board) board certification dates. We had enough information on the Medical Corps to track each of their careers easily through the Navy Medical system. When it came to teaching doctors how to program in NATURAL we all chipped in. Turned out that doctors are really logical thinkers and they adapted to writing their own end user programs like ducks to water. We implemented a Navy Medical wide program to teach every executive to write Natural programs to get the information they needed from the mainframe computer system. I ended up winning an award for writing a program that took their programs and canned them into the "Public" application. This allowed them to share the best of their end user written programs. We soon became familiar with who was the best programmers of the medical, dental, nurse, and MSC corps through this training program and it was my first time teaching about technology. I like it so much I took another part time job teaching driver's education for a few extra dollars at night. I was already coaching the NMDSC guys softball and basketball teams which was fun in itself. The Marines and Navy play an awesome competitive game of basketball in an agency where good health is the primary objective.

27. Dental Corps

The Dental Corps are those medical doctors who specialize in dentistry. Bethesda Naval Medical Hospital had a dentistry building where these cases could be treated. A typical Navy dentist can treat most types of medical problems associated with the teeth, gums, and mouth. The methods used by Navy dentists are state of the art and all equipment is owned and operated by the Navy. At Bethesda Navy Hospital, the president could be a potential patient so they had to be well trained. Many of them retire to private practice and continue serving the nation on a local basis. The Navy also sends other specialties through Dental school such as pilots and others. These dentists have multiple military designators that are valuable to the Navy. They also may retire to private practice. I have known two of these Navy dentists personally who retired and started their own dentistry practice in Maryland. They never knew I controlled their personnel information. My current private dentist is one of these ex-F18 pilots. The corporate database for Navy Medical held all the records for all the dentists in the Dental Corps. It had their education, schools, designators, assignments and much more in arrays of the tables we managed. It was primarily used by BUMED to reassign staff and ensure higher education standards of all dentists. Of primary importance to both doctors and dentists is being Board certified. This was also tracked in the databases. Only the executives at BUMED had access to the data on personnel due to Privacy Act concerns with government data. This information was very important to BUMED and we had control of every dentist's information in the Navy. These were pretty powerful decision making tools for Naval Executives whom we trained in our programming language called "NATURAL". I can't really say that I remember any dentists in our training, but anyone could have been one in a previous assignment and no one except those with access to the personnel data would know. That person was called the "Detailer" and he/she was usually enlisted and controlled the next assignments for every BUMED staff member. He was a good person to know at BUMED if you were up for reassignment.

28. Nurse Corps

There is nothing like an angel by your bedside when you are injured in the hospital. The women of the Navy Medical Nurse Corps are very talented and well trained. Many of them become senior executives at BUMED. They become excellent Health Care Administrators after their nursing careers. We had many whom were trained to use our database software and could write their own programs to extract information from the corporate database. They should not be underestimated. The information tracked on all the Nurses is similar to that tracked on the Doctors, MSCs, Dentists, and enlisted men of Navy Medical. Today, nurses have college degrees and are registered and also certified. These highly professional staff members provide excellent health care in the fleet as well as at Navy Clinics. Many of the nurses teach at nursing school after they have completed their training. The NIH Medical Library has resources to train all the specialties in medicine. This is one avenue for continued education of all staff. The library is also an online resource to medical problems through MEDLINE. It is a 24 our a day operation of online computers and a database that doctors can query from any country in the world. They simply give the patient symptoms and MEDLINE will return the most likely diagnosis for those set of symptoms. Many hard to discover cases have been solved using MEDLINE by doctors in remote regions of the world. The 24 hour access means that doctor has the best library resources in the world at his disposal if he only has a telcom line

80

and computer to query the MEDLINE databank. Now MEDLINE also comes

out on CDROM and is distributed to doctors for subscription fee. This gives

him the power at his fingertips when he needs to ask the questions and get

quick answers to save patients.

Nurses were not allowed on the ships of the past due to restrictions to

men only crews. They were allowed to work at the shore facilities. Some

famous nurses include Clara Barton (civil war), Florence Nightengale, and

others.   They provide the doctors assistance in surgery as well as in non life

threatening situations. They receive the RN designation when they are

certified which stands for Registered Nurse. This can be done through

community college classes and higher levels.   Four year degrees are

common these days and qualify the nurse for higher level responsibilities.

29. Medical Service Corps

The Medical Service Corps are the administrative branch of the Navy Medical Command. The draw their talent from corpsmen and enlisted ranks in naval medicine as well as college. Some of the officers may have wanted to be doctors or physicians assistants but went into administration sciences. They really were a nice lot of guys and most really believed in good health. The ones I worked directly for were trained in database administration, finance, personnel, engineering, and computer operations. I usually trained them the best I could in what my duties and responsibilities included. As a young college graduate my opinion was useful in most cases but advice not always taken except by one MSC officer. He let me train him on how to do database administration the University of Maryland way. I told him about James Martin book series and he ate it up. Next thing I knew he was reading the DBMS Management book as a guide to where we should be going in our DBMS branch. I did not know at the time that his teenage son was attending University of Maryland but now understand why he took my advice since my son attended there also. We were becoming one of the best Corporate Database shops in the government and interacted with all types of Navy and vendor companies. He sent me to the ADABAS International Users Group meeting in Nashville, Tennessee and I was glad to go and bring back the mother load of information for our branch operations. This MSC trusted me like a son and I was never again to know that feeling from any boss I would ever have. He said there would be a chance a civilian DBA leadership position might be created. That turned out to be wishful thinking as the civilians almost never had authority over the MSC's at Navy Medical, except on the rare occasion that the civilian was also a military reservist. At least I knew where I stood which was pretty good. I built several software systems under this MSC and did some of my best work in my career. I was damned glad to have a leader who took my advice and encouraged me to go into management. He even paid for some of my graduate courses at American University which was the Navy Medical school of choice back then for personnel development. It was expensive, but I understood why by the end of my graduate program. After this leader rotated out of the command, I had to pay my own college bills and that was difficult on a GS-12 salary which was $35,000 at the time with two kids. I found out that he had great pull with his boss, another MSC and a navy captain. He liked my work a lot and bragged about me all the time. I introduced him to my father. The DBA MSC knew I would work as hard as I could and man could he motivate me. He served in Vietnam with the Marines and I knew I could trust him. When you work for someone like that, you hope and pray he doesn't move on to a promotion too fast, but that was just what happened. He made me employee of the week twice. He repaid my loyalty to him in a cash bonus when he left our office. He had me thinking I could leave federal service and become more than just a technical guru. In fact, I ended up working for Software AG in Reston, Virginia for two weeks which I hated because I was not in direct contact with management and did not have the authority I had earned at Navy Medical under the MSC officers.

I also learned a few business tricks from my MSC. He taught me how to check on a person's background education before you meet with them. He taught me how to meet and direct my attention to the goals and objectives of the organization. He sent me to work on some committees that I still remember today. His replacement was a nice guy too and had me doing things for my career like learning contracting on the TRIMIS project. I became an evaluator on the TRIMIS project with other civil servants and learned directly about evaluating a large contract. I would later go to contract management school at George Washington University.

30. Civilians

Civilians are the backbone of the technical force supporting the Navy Medical Department. Without these employees there would be no continuous service and operation of shore based operations. The military is the only organization in the United States where the civilians run the politics and are run by the military officers in the service. Civilians are many times ex-service men and relatives of service men themselves. Data Processing, Research, Statistics, and Administration positions are all better accomplished by civilian employees who can dedicate a career to the Navy in one position. Rotating positions are usually reserved for military officers and enlisted staff although some civilians can rotate through various civilian positions in the US Civil Service. The pay rates for civil service employees are reasonable but the pension system has been weakened since the 1980's. Many civilians in the Navy Department have contributed to the overall success of the department. Civilian employees may rise to the level of Secretary of the Navy. The Surgeon General of the United States is a civilian. The Surgeon General of the Navy is usually a military officer. BUMED has many captains. The Pentagon has many officers and civilians who work with BUMED to support Naval Operations. Field level civilians are usually technicians whereas Pentagon level civilian employees are political appointees. The civil service grades range from GS-1 to GS-15 and the Senior Executive Service levels and jobs are found through the OPM (USAJOBS) and Navy Personnel Office.

31. MTF's – Navy Hospitals and Clinics

The Navy calls all medical locations around the globe "Medical Treatment Facilities" or MTF's. This includes 300-1000 bed hospitals, clinics, and other locations at Naval Bases and Marine Corps Bases. In our data processing operation we would typically keep accounting records of operational costs at all the MTF's and the statistics of inpatient and outpatient visits at the MTF's by MTF name. The accounting system kept enough information to let BUMED executives make the operation more efficient in terms of costs. This data was the lifeblood of the management of Navy Medical as a command. We could regionalize the data and group MTF's in their fleet locations around the world. We could pull data through our interactive computer interface on a specific MTF. The total number of MTF's included mostly shore based operations. I never remember any processing of hospital ship data but they may have been included as MTF's and I did not realize it because I never set foot on one at the time even though the USNS Comfort was harbored in Baltimore. Many of the Navy Medical staff had served on the Comfort at some point in there career and it would have been an honor they would have chosen to do for one assignment if they were the right designator. The navy clinic I visited for a systems project was Pax River. The navy hospitals I did work at included Portsmouth Naval Hospital and Bethesda Naval Hospital. I had been a dependent patient at Bethesda and Pax River. I trained staff at Portsmouth on a new computer system we installed for them in the Atlantic region. During this time I was also evaluating civilian hospitals in the Washington area for possible future jobs and I was comparing services of the MTF's to services offered by local civilian hospitals where I had been a patient. The navy seemed to have their own MTF world. The civilian hospitals did pay more for the same services but I never chose to work for any. One learns that there are teaching hospitals like Bethesda with USHS and NIH and there are field hospitals like Okinawa, San Diego, Pearl Harbor, Italy, etc. where the field naval bases are located. Clinics could do any medical procedures that a naval hospital could do in an emergency. But usually major surgery was done only at the hospitals because that is where the major medical equipment and recovery rooms and more experienced doctors are located. My experience with Air Force hospitals helped me realize that Naval hospitals were very advanced in comparison to field hospitals. The list of all navy MTF's is located in appendix A.

32. Regions

The military is often very hierarchical by nature and the formation of regions in the organization charts is probably common to most American forces. CONUS designates continental US. Pacific Region includes Hawaii and Japan. Atlantic region includes the Atlantic Ocean Region. Northwest indicates the northwest United States. The regions are global in coverage as are the navy hospitals themselves and the navy. There are more navy hospitals overseas than any other types from America and they serve all American servicemen and women.

33. NMRI / NMRC

The following is an extract from the NMRI webpage. NMRI is based at Naval Hospital Bethesda and I had several friends who worked there and told me it was the one of the leading places for research in Navy Medicine in the world and that they consorted with NIH when they could on certain cases. They sent their staff people to George Washington University which has an outstanding medical school. The NMRC mission statement[6] is

"The NMRC mission is to conduct research, development, tests and

evaluations to enhance the health, safety and readiness of Navy and Marine

Corps personnel in the effective performance of peacetime and contingency

missions, and to perform such other functions or tasks as may be directed by

higher authority".

The history of the NMRC is outlined on their webpage as follows:

"The Naval Medical Research Center (NMRC) was established in 1942 as the Naval Medical Research Institute (NMRI) on the grounds of the National Naval Medical Center in Bethesda, Maryland.

During World War II, NMRI research investigations were related

---

[6] NMRC webpage accessed on 29 July at www.nmrc.navy.mil

to immediate military operational problems. Scientific investigations in this period included efforts such as: the development of repellants to counter shark-attacks on deployed Navy personnel such as divers; the development of repellants against insect vectors of disease in remote geographical locations; heat stress studies that led to exposure limits for hot-humid shipboard environments; and various investigations for the development of safety equipment including protective clothing, flight goggles, and safety belts. NMRI investigators were commissioned to study the Japanese survivors of the atomic bomb and became deeply involved in developing methods for use in treatment of radiation exposure. These efforts led to the establishment of the Armed Forces Radiobiology Research Institute.

During the 1950's and 1960's NMRI became involved in training monkeys, and later human astronauts, for space flight. These involvements led to the development of telemetry for transmitting astronaut physiological data, such as heart rate and blood pressure, from air to ground.

Another critical NMRI mark of accomplishment during this period was the establishment of the Navy Tissue Bank where NMRI investigators developed freeze-drying techniques for tissue preservation for grafting and other reconstructive surgical procedures.

During the Vietnam War thousands of wounded soldiers were treated with tissue that was collected, preserved and shipped from the Navy Tissue Bank.

Over the years, Navy researchers at NMRI and then at NMRC have pioneered unparalleled scientific advances in a wide variety of cross-disciplinary fields of research and development. Navy researchers have distinguished themselves in investigations regarding: the treatment of septic shock and shock due to blood loss; the development of heart-lung machines; highly efficacious methodologies to prepare and protect patients on the operating table by developing a technique to lower patients' body temperatures for more successful open heart surgery; studies of survival and resistance training; and, frostbite therapy. The Navy environmental stress research continues to this day to enhance the safety and efficiency of Naval personnel performing in hazardous environments.

In the mid-1970's the construction of a multi-million dollar Hyperbaric Research Facility was begun and commissioned in July 1981. It featured five connecting dry pressure chambers and a single water chamber all of which are capable of being set at various pressure levels.

As a result of a variety of important initiatives, by the mid to late 1980's the Navy became a singular research leader for the Department of Defense Marrow Donor Program (C.W. Bill Young Marrow Donor Recruitment and Research Program). NMRI's bone marrow research efforts over the years have provided military contingency support for casualties with marrow toxic injury due to radiation or chemical warfare agents. Biomedical scientific personnel continue to perform laboratory research which supports technology innovations to make highly reliable and cost effective DNA-based typing for marrow transplants.

In the 1990's NMRI research successes continued and singular honors were bestowed upon NMRI investigators. On 2 February 1995 the Space Shuttle Discovery lifted off during a spectacular launch from the Kennedy Space Center on a nine day mission. The payload included an experiment developed by NMRI scientists who were investigating the growth and development of bone marrow stem cells. On 7 September 1995 the Endeavor was launched into Earth Orbit with the second set of experiments developed by NMRI scientists.

In this same period, NMRI investigators made major and unprecedented strides in whole organ transplant procedures using immune modulating agents thus avoiding the necessary use of lifelong immuno-suppressants. This success resulted in a major collaborative effort between the Navy and the National Institutes of Health.

Also in the 1990's NMRI investigators assumed distinguished and widely acclaimed scientific leadership roles in the areas of infectious diseases and biological defense. In 1995 NMRI investigators developed a diagnostic assay for the rapid detection of biological threat agents that was chosen by the Combat Developer over all Department of Defense medical research methodologies for advanced development.

By the mid 90's and toward the end of the decade, NMRI investigators had embarked upon highly acclaimed research methodologies in infectious diseases leading the way to innovative basic investigations and efforts for unique vaccine development against mission abortive problems such as malaria, enteric pathogens and dengue viral fever. It is no wonder that by the end of the 1990's, Navy investigators merited high honors such as the Frank Brown Berry Award and the Conrad Dexter Award.

On 01 October 1998, the Naval Medical Research Institute was disestablished and succeeded itself as an Echelon 3 headquarters under its new name, the Naval Medical Research Center. As an Echelon 3 headquarters, NMRC was missioned to provide leadership for the Navy's biodental research laboratory at Great Lakes, Il, and for the

Navy's overseas laboratories in Peru, Indonesia and Egypt.

In June 1999, along with the Walter Reed Army Institute of Research, NMRC moved to the new federal research laboratory facility erected in the Forest Glen section of Silver Spring, Maryland.

Today, NMRC research continues to unfold with ever increasing success and unquestioned high acclaim at NMRC itself, in other Navy laboratories, and in partnership with agencies. Non-federal collaborations are promoted through an extremely successful and active technology transfer program that includes various cooperative research and development agreements (CRADA's) with universities and private industries. Navy-supported medical research efforts have influenced the civilian practice of medicine, assisted the Ministries of Health in developing nations, and provided technology for other federal initiatives."

Appendix A. Navy Medical MTFs

Medical Treatment Facilities

Click on the Site you Wish to Visit:

Guam   **Naples, Italy**   **Rota, Spain**

Guantanamo Bay, Cuba   **Okinawa, Japan**   **Sigonella, Italy**

Keflavik, Iceland   **Pearl Harbor, HI**   **Yokosuka, Japan**

London, United Kingdom **Roosevelt Roads, Puerto Rico**

Appendix B. Tri Military Medical Services Centers

Participating Military Medical Centers

- Brooke Army Medical Center, San Antonio, Texas

- Madigan Army Medical Center, Tacoma, Washington

- Malcolm Grow Air Force Medical Center, Andrews AFB, Maryland

- National Naval Medical Center, Bethesda, Maryland

- Naval Medical Center, Portsmouth, Virginia

- Walter Reed Army Medical Center, Washington, DC

- Wilford Hall Air Force Medical Center, Lackland AFB, Texas

- Eisenhower Army Medical Center, Ft. Gordon, Georgia

- Naval Medical Center, San Diego, California

Imperial War Museum

Department of Printed Books

Recommended Reading List (No. 353)

Royal Navy Medicine

This is not a complete bibliography but a selected list of items that should prove particularly useful and be relatively widely available. You are welcome to make an appointment to consult them in our Reading Room, where you can also study our catalogues for further references. The Reading Room is open between 10am and 5pm Monday to Friday and the same hours on most Saturdays, although this is a more limited service.

If you are unable to visit the Reading Room, many items in this list should be available through the inter-library loan scheme. Your local public library should be able to provide more details. Please note that, as a national *reference* library, our stock is not available for loan. It may be possible to purchase some of the books in this list *but please note that none is sold by the Department of Printed Books unless a note to that effect is given*. Once books are out of print they may be difficult to track down and could be expensive. We have a list of second-hand military book dealers who may be able to locate them. You could also try *www.clique.co.uk* on the Web, where you will find details of over 1,350 British-based dealers and links to an international database of many millions of second-hand books.

The Department of Printed Books has reprinted some key works for sale, but these represent only a tiny proportion of the books held in our library. A catalogue of our publications for sale is available on request or can be downloaded from our website.

If copyright law and conservation considerations allow, we may be able to provide photocopies of journal articles or extracts from published books. Prepayment is required and you will need to complete our copyright declaration/request form for each item. When requesting copies, please state in which reading list the item was found and give full details of the item, including our classification number if appropriate and the page numbers required.

In addition to Printed Books, the Museum has six other Collecting Departments – Art, Documents, Exhibits and Firearms, Film & Video, Photographs and Sound. These may hold further information about this subject. All can be visited by appointment. A leaflet about the Collecting Departments is available on request.

Department of Printed Books, Imperial War Museum, Lambeth Road, London SE1 6HZ. Tel 020 7416 5342 Fax 020 7416 5246 E-mail books@iwm.org.uk Website www.iwm.org.uk

Allison E Duffield / 04.01.02

ADMIRALTY

Handbook of the Royal Naval Sick Berth Staff / [Admiralty]. -
London : HMSO, 1944. - ix, 526p., 4p. of plates (one fold.): ill.
(some col.) ; 25cm. - index. (B.R. 888).
Our Classification: 81(41).02/5-0   Our Accession No.:   58405

ADMIRALTY

Instructions for the Royal Naval hospitals and other medical
establishments at home and abroad, 1927 / [Admiralty]. - London :
HMSO, 1927. - iv, 276p. ; 25cm. - index.
Our Classification: 13(41).81/4-0   Our Accession No.:   01 / 269

ALLISON, R.S.

The surgeon probationers / R.S. Allison. - Belfast : Blackstaff,
1979. - xviii, 142p.: ill., facsim., 4 ports. ; 22cm. - index.
ISBN 0-85640-145-5
Our Classification: 13(41).81/3   Our Accession No.:   79 / 3686

AVERY, K.R.

A dose of salts / Lieut. Commander K.R. Avery, RN (Retd). -
Braunton, Devon : Merlin Books, 1988. - 270p.: ill., facsims., ports.
; 21cm.
ISBN 0-86303-380-6 (pbk.)

| Our Classification: | 23(=41)/1 [Avery, Kenneth R.] | Our Accession No.: | 88 / 2458 |

BAGSHAW, ROBERT

Toothy goes to war: memoirs of a dental officer 1939-46 / by
Robert Bagshaw. - Wymondham, Norfolk : George R. Reeve, 1987.
- 175p.: ill., facsims., frontis., ports. ; 21cm.

ISBN 0-900616-25-3 (pbk.)

| Our Classification: | 23(=41)/5 [Bagshaw, Robert] | Our Accession No.: | 88 / 187 |

CASUALTIES AND MEDICAL STATISTICS

Casualties and medical statistics / edited by W. Franklin Mellor. -
London : HMSO, 1972. - xiv, 893p. ; 26cm. - index. (History of the
Second World War : United Kingdom medical series).

ISBN 0-11-320997-5

Our Classification: 94.1(41)/5   Our Accession No.:   67659

CLARK, GREGORY

"Doc": 100 year history of the Sick Berth Branch / Gregory Clark. -
London : HMSO, 1984. - 172p., 16p. of plates: ill., facsims. ; 22cm.
- bibl. p.168 - index.

ISBN 0-11-290427-0 (pbk.)

Our	13(41).8181 [Royal Naval Sick	Our Accession	85 /
Classification:	Berth Branch]/1	No.:	544

COULTER, J.L.S.

The Royal Naval medical service / by Surgeon Commander J.L.S.
Coulter. - London : HMSO, 1954-1956. - 2 vols.: ill., map ; 25cm. -
index. (History of the Second World War : United Kingdom
medical series).

Our Classification: 13(41).81/5 Our Accession No.: 33117

DRURY, JOYCE

"We were there" / Joyce Drury. - Dudley, West Midlands : Jupiter
Press, 1997. - 181p.: ill., figs., ports. ; 21cm.

ISBN 0-9530620-0-7 (pbk.)

Our	38(41).23 [Voluntary Aid	Our Accession	00 /
Classification:	Detachments]/5-2	No.:	1668

FIRST AID IN THE ROYAL NAVY

First aid in the Royal Navy / [Admiralty]. - London : HMSO, 1943.
- 106p., 1 fold. leaf of plates: ill. ; 18cm. (B.R. 25).

Our Classification: 81(41).488/5 Our Accession No.: 18830

FIRST AID IN THE ROYAL NAVY

First aid in the Royal Navy - London : HMSO, 1914. - 190p., 2

fold. leaves of plates: ill. (some col.) ; 14cm. - index.

Our Classification: 81(41).488/3   Our Accession No.:   699

HARLAND, KATHLEEN

A history of Queen Alexandra's Royal Naval Nursing Service /

Kathleen Harland. - N.p. : Journal of the Royal Naval Medical

Service, [1990]. - 169p., 18p. of plates: ill., facsims., ports. ; 21cm.

ISBN 0-9514906-0-5 (pbk.)

Our Classification:	13(41).8181 [Queen Alexandra's Royal Naval Nursing Service]/1	Our Accession No.:	90 / 1015

HORTON, GERALDINE

No nightingale sung / Geraldine Horton. - Hythe, Kent : Volturna

Press, 1996. - viii, 342p.: figs. ; 20cm.

ISBN 0-85606-361-5 (pbk.)

Our Classification:	23(=41)/5 [Horton, Geraldine]	Our Accession No.:	98 / 2200

JOLLY, RICK

The red and green life machine: a diary of the Falklands Field

Hospital / Rick Jolly. - London : Century Publishing, 1983. - xii,

146p., 24p. of plates: ill. (some col.), ports. ; 24cm.

ISBN 0-7126-0158-9

Our Classification: 23(=41)/7 [Jolly, Rick]   Our Accession No.:   83 / 1421

MCBRYDE, BRENDA

Quiet heroines: nurses of the Second World War / Brenda

McBryde. - London : Chatto and Windus, 1985. - x, 246p., 4p. of

plates: ill., ports. ; 23cm. - bibl. p.240-241 - index.

ISBN 0-7011-2939-5

Our Classification: 81(41).42/5   Our Accession No.:   85 / 1646

REVELL, A.L.

Haslar. The Royal Hospital / by A.L. Revell. - Gosport, Hants :

Gosport Society, [1979]. - vi, 50p.: ill., plans, ports. ; 26cm. - bibl.

p.48 - index.

Our Classification:	13(41).811 [Royal Naval Hospital, Haslar]/1	Our Accession No.:	K. 81 / 348

WALLIS, R. RANSOME

Two red stripes: a naval surgeon at war / R. Ransome Wallis DSC, MD. - London : Ian Allan, 1973. - 144p., 16p. of plates: ill., maps ; 24cm.

ISBN 0-7110-0461-7

Our Classification:	23(=41)/5 [Wallis, R. Ransome]	Our Accession No.:	79 / 3930

WHITE, J.R.S.

H.M.S. Norfolk and other ships, 1940-1946: the wartime diary of a dental officer / by Surg. Lieut. Cdr. (D) J.R.S. White, RNVR. - Leeds : Fractal Press, 1995. - 103p.: ill., map, ports. ; 21cm. - index.

ISBN 1-87073-505-6 (pbk.)

Our Classification:	23(=41)/5 [White, J.R. Stuart]-2	Our Accession No.:	96 / 2186

WILLIAMS, EVE

Ladies without lamps / Eve Williams; illustrated by Mary Ellis. - London : Harmsworth, 1983. - 178p.: ill. ; 23cm.

ISBN 0-9506012-5-X

Our Classification:	23(=41)/5 [Williams, Eve]	Our Accession No.:	84 / 2128

INSTITUTE OF NAVAL MEDICINE

Welcome to the Institute of Naval Medicine! This Royal Navy shore establishment at Alverstoke in Gosport has developed over the last 50 years to be a nationally and internationally recognised centre of excellence for occupational health advice, information,training and research.

RECENT, CURRENT AND FUTURE WORK OF THE INSTITUTE OF NAVAL MEDICINE

Objectives ; Customers ; Organisation ; Courses

Organisation of INM

There are 130 medical doctors, scientists, trainers and health administration staff in the Institute team working in seven Divisions;

Occupational and Environmental Science

Environmental Medicine

Radiation and Submarine Medicine / Diving and Hyperbaric Medicine

Training

Medical Administration

Statistics

Corporate Services

The team is dedicated to excelling in achieving our Mission and giving our customers best value for money for the services they need.

The Mission of the Institute of Naval Medicine is to improve the operational capability of the Royal Navy by promoting good health and safety and maximising the effectiveness of personnel.

The Institute is accredited as an Investor in People, a measure of our commitment to ensuring that staff are appropriately informed and trained for their tasks and responsibilities and have ample opportunities for training and education to help them to fulfil their longer term potential.

Our five principal business areas are:

- *Scientific advice on maritime and military health and safety* based on wide expert knowledge, long experience and an international network of expert colleagues.

- *Operationally deployable specialist medical and scientific staff* providing consultant based preventive and clinical services principally in diving, submarine and radiation medicine.
- *Specialist training* policy and delivery in naval medicine, safety and radiological protection - provided principally by on-site specialists.
- *Research and equipment testing* where the knowledge required to give advice is not available. Much of this work is done in close association with colleagues in the Defence Science and Technology Laboratory (Dstl), universities and allied navies.
- *Corporate services* for the Royal Naval Medical Service including medico legal advice, medical resettlement, libraries and biostatistics, and conference facilities.

Resources and Customers

The Institute of Naval Medicine is unique in maritime and general occupational health in the United Kingdom in the range of specialised facilities and staff from appropriate disciplines working together on a single site. Our most recent major investment has been integration and modernisation of Occupational and Environmental Science Division's laboratory services to provide the high standards essential to occupational and environmental health, especially in the submarine service.

We are on target for National Charter Mark for the Medical Records Department and UK Accreditation Service certification of the Occupational and Environmental Science laboratories.

While most work is in support of the Fleet, with much focused on the Royal Marines and Submarine Service, tasks are undertaken for other authorities and Agencies of the Navy Department and for the Army, RAF, and QinetiQ and other commercial organisations.

Much of the strength of the Institute lies in knowledge, skill, experience, hard work and flexibility of its staff; the collaborative and synergistic relationships between its Divisions and externally as we supplement in-house expertise with deep specialists from universities, and especially our close links with our Customer and the efforts taken to appreciate their requirements. To that end medical and scientific staff spend much time with Service operational and command staff and in the field developing task specifications and gaining experience of the working environment to ensure that aims, requirements, service conditions and constraints are fully appreciated. Liaison continues after the work has been delivered to check that the output meets requirements and fulfils a need.

The high volume of quality outputs achieved reflects the enthusiasm and application of the staff to meet and resolve challenges to the health, safety and personnel effectiveness of the Royal Navy. Some of their reports and publications are listed on this site.

Ethics and care of research volunteers

Research proposals are reviewed and the work authorised and overseen by the MOD(N) Personnel Research Ethics Committee, reporting to the Director of Science (Sea), and the Medical Research Council supported Scientific Advisory Group for INM which reports to the Medical Director General (Naval).

The Institute is especially grateful to the volunteer subjects who take part in our many trials. Operationally important questions would remain unanswered without the contribution of these men and women - and the encouragement and co-operation of those who release them from their normal duties. The interests of volunteer subjects are paramount in my management of research. Great care is taken to make trials and experiments safe. Should something go wrong and they are damaged "no fault" compensation arrangements are in place.

Institute staff lead or contribute to health committees and panels in NATO and with other allied countries. Where security considerations permit, staff are strongly encouraged to submit the results of their work to the scrutiny of the international scientific and medical community through papers given at conferences or publications in the open literature.

Outputs

Specialist scientific and occupational medicine services and advice (backed by research where necessary) and training in maritime medicine and radiological protection are the Institute's stock in trade.

As a general indication of the level of activity over the year 16 new research protocols were approved, 52 scientific reports were published, a consultant opinion was given on the fitness to dive or undergo submarine escape training of 164 people, 395 people were considered by the Medical Board of Survey, the Duty Diving Medical Officer responded to 419 telephone requests for advice on diving or hyperbaric medicine, 813 students attended 72 courses, 6000 samples were processed by the chemical laboratory, and the medical records of some 13000 ex-Service personnel were processed and many thousands of associated written and telephone enquiries answered.

Highlights of advisory, clinical, research and new training outputs have included:

- Close involvement in development of RN Fitness Test policy.
- First phases in development of a Fitness Test which relates to essential work at sea.
- Identification of an appropriate safety helmet for boat crews.

- Introduction of Royal Marine Recruit injury database and work with Commando Training Centre RM to reduce injury rates and severity and speed recovery.
- Policy and actions to reduce heat strain and burns, especially in firefighters.
- Progress towards specification of rehydration fluids and regime for troops during and after exercise.
- User assessments of the proposed Naval Action Clothing System.
- Specialist opinions on the employment of women as submariners and as clearance divers.
- Progress toward better survival clothing and rations for submariners.
- Rationalisation and specification of immersion suits.
- Initial External Validation of Medical Assistant training.
- New course on Radiation Medicine for Medical Officers.
- Proposals for continuation training for medical branch ratings.
- Review of the pathophysiology and treatment of underwater blast injury.
- Hyperbaric therapy caseload increases and joint research studies with the Royal Marsden Hospital.

Publications and presentations

Surviving peer review is a good test of the quality of the Institute's work and raises the profile of the Royal Navy. Over the last two years presentation and publication of research results have been strongly encouraged and some time has been made available in study programmes for their preparation. There has been a marked increase in the number of papers contributed to conferences and in the papers and abstracts which have been published in the scientific literature. This reflects well on the skill and application of staff.

Heritage

The Institute has grown from a modest beginning when it was housed in Monckton House, a fine mid-19th century gentleman's mansion which now accommodates the management centre and the Royal Naval Medical Service Historic Library and Collection, comprising books and other documents from the libraries and museums previously housed at the Royal Naval Hospitals Plymouth and Haslar. This Library and Collection is maintained with the assistance of the Hampshire County Council which has kindly agreed to catalogue and conserve the collection of documents and photographs and that work is underway. It is also to provide a part-time specialist librarian and survey the museum artefacts to advise on their display.

The Queen Alexandra's Royal Naval Nursing Service Musuem and Archives is also accommodated in Monckton House.

An affiliation has been forged with the Worshipful Company of Barber Surgeons to promote the study of the health of the Fleet and history of maritime medicine.

The way ahead

Knowledge that the Institute of Naval Medicine will have a key role in the re-organised Defence Medical Services post-SDR and in promoting the health and effectiveness of the Royal Navy beyond the Millennium, catalysed by our commitment to "Investors in People", provided a firm foundation for constructive change, purposeful strategic planning for outputs and development, improved management of training and better internal communication.

Our order book of work is brimming over for next year and some years beyond as interest increases in the benefits to be gained by achieving high standards of health, safety and environmental protection and improving workplace design to increase the effectiveness of personnel. We intend to provide the quality and speed of response required by the Royal Navy so that it may always be ready "To fight and win" and by our many other Customers. We look forward to developing strong links with the Centre for Defence Medicine.

Appendix E: The Role of Royal Navy Medical Service

The Primary Purpose of the Medical Director General (Naval) is to:

'Conserve manpower by ensuring that the Royal Navy has a medical capability for war which meets the standards set by the Surgeon General and is adequate to maintain the operational effectiveness, availability, endurance and health of Naval personnel in peace, crisis, major crisis and war.' This purpose has remained constant over many years in the face of rationalisation and reorganisation of the Navy as a whole and the Medical Services in particular. In 0addition, medical science has advanced at an unprecedented rate and the expectations of patients have, quite rightly, increased accordingly.

OPERATIONAL SUPPORT

Medical personnel are deployed in all arms of the Royal Navy; the Surface and Submarine Flotillas, the Fleet Air Arm and the Royal Marines. In addition to their medical training they receive special to arms training. Operational Support is delivered in three stages of increasing sophistication;

First Line Support

At first line, Medical Assistants in ships provide day-to-day medical support and preventive medicine often in isolation from a Medical Officer. In preparation for conflict, they train non-medical personnel from within the ship's company to form first aid parties who, during action, will recover injured personnel rendering emergency first aid in preparation for evacuation. Medical Assistants with Commando Units assist the Medical Officer in the provision of routine medical support. During fighting, they man Regimental Aid Posts treating and holding casualties until they can be moved away from the front line. Medical Assistants undergo Commando Training and earn the right to wear the coveted green beret. In peace, Medical Officers are appointed to look after the medical needs of personnel within a group of ships. Although based in one ship, the Squadron Medical Officer may be required to answer an emergency occurring anywhere within the group. In war, it is planned that all Destroyers/Frigates and above will have a Medical Officer to ensure rapid and aggressive resuscitation of casualties. There is also a requirement to enhance most Ships Taken Up From Trade (STUFT) with a medical capability Dental primary care, in peace, is achieved by deploying Dental Officers to RM units, aircraft carriers and some amphibious ships, with occasional temporary mobile teams deployed to such areas as the ARMILLA patrol. In war, such care would be enhanced with mobile teams at the rate of one team per several Destroyers/Frigates. They would probably be deployed to accompanying RFAs. The Medical Service fully meets the requirements for first and second line units. At third line the situation is more complex.

Second Line Support

In peace, an emergency surgical capability is maintained in aircraft carriers which, coupled with the indigenous helicopter support, provides a substantial medical asset to the Carrier Group. For crisis, additional surgical teams are deployed to the Medical Squadron of the Commando Logistic Regiment, 3 Commando Brigade and into certain amphibious or Royal Fleet Auxiliary Ships.

Third Line Support

Third line support comprises substantial surgical facilities in Primary Casualty Reception Ships. RFA ARGUS has a permanent hospital unit on board and can be manned at short notice. Other hospital units can be built into Ships Taken Up From Trade (STUFT), or other assets, to meet the requirements of specific operations. Depending on the distance of the amphibious operating area from the Forward Logistic site there will be a need for a carousel of casualty transport to evacuate personnel to a suitable airhead for movement to the UK.

CURRENT CAPABILITY

The Medical Service fully meets the requirements for first and second line units. In addition to the permanent surgical teams in HMS ILLUSTRIOUS and HMS INVINCIBLF, Naval Surgeons are regularly deployed to Bosnia, Cyprus, Gibraltar and the Falkland Islands. At third line the situation is more complex. The Royal Naval Medical Service can man RFA ARGUS, as a Primary Casualty Reception Ship, but there are shortages in some specialties, notably; surgeons, anaesthetists, operating theatre assistants and some specialist nurses. Not surprisingly, these shortages mirror both national shortages and shortages evident in the other Services. It has also been decided that for certain specialities there will be a lead Service; Laboratory Technicians will all be RAF. Apart from the Lead Service posts, the Navy aims to combat this deficiency by taking measures to improve retention and directing training into these shortage categories. In the meantime, third line support will need to rely on support from Royal Naval Reserves and QARNNS(R) (RNR & Individual) and, in dire crisis, some specialties from the Army and Royal Air Force Medical Services.

THE MEDICAL AND DENTAL AGENCIES

The Defence Secondary Care Agency, Defence Dental Agency and the Medical Supplies Agency formed in 1995. This change has been particularly evident in the secondary care area with the closure of all but one UK based Service hospital and the formation of Ministry of Defence Hospital Units within NHS Trust Hospitals. A substantial expansion programme to meet the needs of the three Services continues in the Royal Hospital Haslar. With all these changes taking place the Secondary Care Agency inherited a large backlog of patients. This backlog has been tackled through a number of initiatives and a four week waiting time for first appointment is now in sight in many specialities. The concentration of assets into a single core hospital and the formation of partnerships with NHS Trust Hospitals is a realistic method of providing secondary care support to the standard demanded by today's Service personnel. These changes have been difficult and we have asked, and continue to ask, a great deal of our staff. The flexibility and good humour shown, particularly by junior personnel, is something of which we can all be justifiably proud.

Acronyms and Abbreviations

ADM – Admiral

AM – Aerospace Medicine

Limey – derogatory name for a British Sailor from using lime juice to prevent

scurvy

BUMED – Bureau of Medicine and Surgery

BUMED Code – Office at BUMED Headquarters

BUMIS – BUMED Information System (personnel)

CPT – Captain

CLIN – Naval Clinic

COMM – Commodore or short for communications

Corpsman – Enlisted Field Doctor

CPO – Chief Petty Officer

DC – Dental Corps

DCRT – NIH Computer Facility

DEPSECNAV – Deputy Secretary of Navy

Doc – Short for Doctor, Medical Doctor

DOD – Department of Defense

DODCI – Department of Defense Computer Institute

Echelon – Another Military Organization at a different authority level

EMS – Emergency Medical Services

ENS – Ensign

Epidemiology – Medical Statistics

ER – Emergency Room

FS – Flight Surgeon

ICU – Intensive Care Unit

IP – Inpatients

HCMIS – Health Care Management Information System

Hippocratic Oath – First do no harm

LCDR – Lt. Commander

LT – Lieutenant

LTJG – Lieutenant Junior Grade

MD – Medical Doctor

MIPR – Medical Information Processing Report

MSC - Medical Service Corps

MTF – Medical Treatment Facilities

MUMPS – Medical Computer System Language

NAVAIR – Naval Air Command

NAVHOSP – Naval Hospital

NAVINSTR – Naval Instruction (memo from higher echelons)

NAVMEDCOM – Naval Medical Command

NAVSUP – Naval Supply Command

NAVREG – Naval Region

NC – Nurse Corps

NIH – National Institutes of Health

NLM – National Library of Medicine

NMRI – Naval Medical Research Institute

OP – Out Patients

OPNAV – Director of Naval Operations

PA – Physicians Assistant

PEB – Physical Evaluation Board

PM – Project Management

PP – Plant Property

PO – Petty Officer

RLG & RDT - General Ledger Accounts System

SECDEF – Secretary of Defense

Statistics – Lies and damn lies

TRIMIS – Tri – Medical Information System (Army, Navy, Air Force)

UCA – Uniform Chart of Accounts

USHS – Uniformed School of Health Sciences

USNA – United States Naval Academy

USNS Comfort – Hospital Ship Atlantic Fleet

USNS Mercy – Hospital Ship Pacific Fleet

WWIP – World Wide Inpatients

WWOP – World Wide Outpatients

Bibliography

Books

Chernow, Ron, <u>Alexander Hamilton</u>, Random House, New York, 1984.

Clancy, Tom, <u>Submarine : A Guided Tour Inside a Nuclear Warship</u>, Berkley
Books, New York, 1993.

Carter, Jimmy, <u>Faith</u>, New York, 1999.

Ennes James, CDR, <u>Assault on the Liberty,</u> Random House, New York, 1979.

Herman, Jan K., <u>Battlefield Sick Bay: The History of Navy Medicine in World
War II</u>

Langley, Harold, <u>The History of Early US Navy Medicine</u>, New York, 2001.

McCain, John, Senator, <u>Faith of My Fathers</u>

Montor, Karel and Maj. Ciotti, Anthony, <u>Fundamentals of Naval Leadership</u>,
Naval Institute Press, Annapolis, Maryland, 1984.

Nagle, James , History of Government Contracting, George Washington University Press, Washington DC, 1992.

Nelson, Pete, Left for Dead: The Story of the USS Indianapolis, Random House, New York, 2003.

Stanton, Doug, In Harm's Way: The Sinking of the USS Indianapolis and the Extraordinary Story of it's Survivors, St. Martin's Press, New York, 2001.

Stavridis, James, CDR, Division Officer's Guide, 10th Edition, Naval Institute Press, Annapolis, Maryland, 1995.

US Army, US Special Forces Medical Handbook, 1982.

USNI book club offerings on military medicine

William H. Gray, Gray's Anatomy, Houghton-Mifflin, New York, 1993.

Journals/Magazines

New England Journal of Medicine

Navy Directive 5000 Series – Systems Life Cycle Management

NAVDAC Pubs – Publications for Navy Data Processing / IT

Chips Ahoy – Norfolk Naval Station Computer Magazine

GPO, DFARS – Defense Federal Acquisition Regulations, Title 47 CFR

GPO, FAR – Federal Acquisition Regulations, Title 48 CFR

Knauss, Chris, "Floating Comfort; Navy hospital ship ready for deployment

around the globe", in The Mariner Magazine, 27 May 2005,

NIH, National Library of Medicine Medline

National Contract Management Magazine

Ruff, LCDR, Article on "Navy Nurses in Iraq", Proceedings Magazine, US

Naval Institute, US Naval Academy, 2005

Videotapes, Movies, and Cable TV

Videotape "Victory at Sea" World War II

History Channel Specials on World War II, World War I, Ancient Wars

Movie "Troy", Trojan War, Illiad by Homer

Movie "Mr. Roberts" – World War II, US Navy

Movie "MASH" and TV Series - Korea

TV Series "China Beach" – Vietnam, US Army MASH

Movie – "Top Gun" – Post Vietnam, US NavAir

Movie – "Sands of Iwo Jima" – John Wayne, US Marines

Movie – "Flight of the Intruder" – Vietnam, US NavAir

Movie – "Patton", 1970, World War II, US Army

Movie – "Platoon", Vietnam, US Army

Movie – "Dear Hunter", 1979 – Vietnam, US Army

Movie – "Green Berets", 1968 – Vietnam, US Army Special Forces

Movie – "12 O'clock High", MGM, Gregory Peck , US Air Force

History Channel Series Series – Mail Call on History Channel – GSgt Lee Irmy

Movie – "Windtalkers", World War II, US Marines

Movie – "Midway", World War II, US Navy

Websites

www.navymedicine.mil

www.navy.mil

www.usnscomfort.mil

www.usnsmercy.mil

www.medline.gov

www.nih.gov

www.cdc.gov

www.jhu.edu

www.usna.edu

www.usafa.edu

www.pentagon.mil

www.nlm.gov

www.ndu.edu

navymedicine.med.navy.mil/bumed

www.navynews.uk.co/youngreaders/news.asp

www.rnreference.mod.uk/09/brdsheet.html

Biography

Donald Joseph Gray Chiarella lives in Elkridge-Hanover in Howard County, Maryland with is wife Mimi, and 4 great children. Born in Kilmarnock, Scotland in 1956 to a US Air Force family, eldest son of Donald Sr. and Margaret. Don is now a MIS Section Chief and Supervisory DBA for Maryland State Highway Administration in the Traffic Safety Analysis Division since 1997. He is in the 1997-2003 editions of Who's Who in Information Technology in America. He holds an independent study Ph.D. from Kennedy-Western University in MIS (2001), an M.S. degree in Technology of Management from American University (1988) (Dean's List), and a B.A. degree in Urban Planning / IFSM from University of Maryland (1979) with the first degree in those two specialties. He is certified in local planning by University of Maryland. He is also certified by George Washington University in Public Contracts Management. He is certified by ICCP in IT management. He is also certified in Computer Security Management by ISACA. He worked for the US Navy Medical Command at the Data Services Center in Bethesda, Maryland for 10 years from 1977-1987 where he gained insight to TRIMIS contracting, medical data processing and requirements, Navy Medical Corporate Data Administration, Navy Medical

Business Systems and Medical Programming, NMDSC Operations Division, Headquarters Executives training, and the Non-Tactical ADP Paperless Ship concept. He has attended the USDA Graduate Mgt School, OPM Training Center, GSA Training Center, National Defense University, Department of Defense Computer Institute, NIH DCRT Computer School, and US Air Force Academy. He is a 1974 Nixon presidential and 1975 Maryland state scholarship winner and has won a college championship in baseball (catcher) in the PIC NAIA Conference in 1976 at St. Mary's College of Maryland. He is ACEP coaching certified and served as a little league commissioner for the Upper Marlboro Boys and Girls Club in basketball in 1987. He won an award for staff mentoring from GSA in 1994. He is a life member of the US Naval Institute, American University Alumni, USAF Academy Association of Graduates, and the Institute for Transportation Engineers, IEEE, ACM, and Mathematical Association of America. He is also a member of two Transportation Research Board committees on highway safety. Don has built many computer systems and written many papers and documents for government systems management and planning. He has studied 7 foreign languages and more than 10 computer languages, 7 operating systems, and 5 databases.

Don volunteers as a Howard County Democratic Chief Voting Judge and docent at the Historic Electronics Museum in Linthicum, Maryland sponsored by Grumman and teaches at Anne Arundel Community College and Aspen University in Denver graduate school online. He has taught over 25 courses and written 10 books over the years, some of which are available at

www.lulu.com/don and www.lulu.com/dchiarella and the Barnes and Noble

webpage. He attends Glen Mar United Methodist in Ellicott City, Maryland and

his neighbors have asked him to run for County Council.

www.ingramcontent.com/pod-product-compliance
Lightning Source LLC
Chambersburg PA
CBHW022004170526
45157CB00003B/1144